Underwater Man

To Bill,

With Best Wishes

Joe Mac

Underwater Man

JOE MacINNES

FOREWORD BY PIERRE ELLIOTT TRUDEAU

DODD, MEAD & COMPANY • NEW YORK

Photo credits

All the photographs
in this book are
the property of Joe MacInnis,
with the exception
of the following:

Page 14: Toronto Telegram
Pages 54, 70: Robert Stenuit
Page 92: A. Moon
Page 126: U.S. Air Force photo
Page 134: National Geographic

Excerpts from Robert Stenuit
"The Deepest Days,"
National Geographic, April 1965;
and Joe MacInnis,
"Diving Beneath Arctic Ice,"
National Geographic, August 1973:
copyright National Geographic Society.
Reprinted with permission.

Contents

Sea Heroes

What novel breed of men
Are they
Who venture beneath the sea?
Today's vikings . . .
Climbers of underwater Everests . . .
Free-fallers of inner space . . .
Who negotiate terms
With ancient ocean rites
Their depth-contoured journey
Carries them naked
Where eon's purple ink
Writes dangerous warnings. . . .

Who are these men
Of yesterday and today
Who discuss tomorrow's explorations
In distant weedless whalehalls. . . .

They are you and I
And the Columbus in us
Continually calls
To
   sunless
     valleys
       of the deep

Foreword

Within our lifetime the development of Scuba equipment has opened a fascinating new element of this planet to our scientific curiosity and sense of adventure.

Although still in his thirties, Dr. Joe MacInnis is still a charter member of the first underwater generation. He found his vocation early in life, and it is one which offers a continuing challenge to his energy and imagination, and to his unusual combination of abilities as athlete, scientist and artist. His work, and that of his fellow researchers, is adding to Canada's knowledge of its water resources, and to mankind's.

Diving with the author is an education as well as an adventure. He obviously enjoys sharing his knowledge, as readers of this book will discover. If they are divers, they will recognise many familiar pleasures. If they are not, they may well be persuaded to take the plunge. Judging by my own experience, they will be glad that they did.

Pierre Elliott Trudeau
Ottawa

For Deborah
who was, and is,
the well spring

The Surface

The sea is holding its breath. No waves break against the beach, nor do ripples caress the calm. All is quiet under a serene night sky.

I walk carefully down to the ghostly place where the sea meets the land. It is a strange marriage this; one partner immovable and solid, the other dynamic and fluid. But there is a third consort here, forgotten because of her invisibility. It is the soft night air, delightfully invading my lungs, making a noble triangle with sea and land. The three partners lie together silently as I sit beside their meeting place.

The water below me is black, yet clear. Its mirror surface reflects the skyroad of constellations overhead. Soft shafts of starlight enter, to be lost in the dappled shadows of the depths. Vague charcoal outlines indicate smooth rocks on the seafloor.

The ocean's depths have attracted my thoughts and energy for two decades, urging me toward their secrets with an allure equal to the sirens of Scylla. Twenty years ago I was first captivated; ten years ago my career and the sea became one. The songs from this dimension were too seductive to be resisted.

Whatever the origins, water, in all its form and flow, has always held infinite mystery. Lakes and streams are deeply fascinating and reflect invitation to new discoveries.

I grew up in a country where water dominates the landscape. Ontario has a quarter of a million lakes and long coastlines on two seas, one salty and one fresh. Its cold waters shimmer history. Waves and foam inspire visions of long silent strokes of Indian paddles and birch bark canoes. Forest shores echo the magic of Kagawong, Temagami, and Kawartha. Summers are paradise.

7

When I was ten years old, I encountered the first event which swept me toward the undersea world. It happened on a calm August day. The afternoon sun was hot to the skin. I walked along a dock, savouring the cool lake in front of me.

At that time, water had only two dimensions—length and breadth. Depth was something only in my imagination; meaningful access was prohibited. Short plunges off this same dock had confirmed this, for all I could see were vague flickers of darkness and light. I knew that my eyes were designed for air and could not accommodate to the elusive underwater shadows.

I held in my hand a device that would change everything, for swaying lightly from my finger tips was a diver's facemask.

I made a shallow descent to the dusky fringes of a wood and stone dock. My new and special window revealed a covey of sunfish skittering through the outlines of old rocks and timber. I swam deeper, tightly holding the oval glass and its rubber skirt to my face. Three dark green bass rose in greeting. I looked up and could scarcely see the surface. I imagined myself to be at the very depths of the lake. I was wrong. I had swum down a mere twelve feet.

I returned again and again to the quiet world of bass and sunfish. I was especially curious about the slow-moving creatures who lived near the bottom. But I could not stay long. I was held back by the cold and the need for air.

One summer, early in the fifties, my parents took me to the Maritimes. From a promontory high over the St. Lawrence River, I gazed eastward and into the open ocean. Cartier and Champlain and their voyages suddenly became more meaningful. They were more than school-book stories. These men had sailed their sea-weak craft westward through these waters.

But neither history nor the endless horizon mattered that day. I was captured by a new sensation—the scent of the sea.

It came in on a cool wind, hinting of salt-stained beaches and wandering sea-weed. It murmured the music of slackening tides and unseen waves far from shores. At this point, my subconscious began to weave the ocean into a cloth of possibilities.

A few years later a parabola of events converged all watery preludes into a single focussing experience. I made my first dive into the sea.

I now look back on those days from the lofty perch given by time. I was on the stern of a dive boat. The Florida shore lay soft on the west. I stood at the centre of my second decade; muscles strong from swimming and nervous system open and receptive. I trembled with expectation and unvoiced fear. A few feet below lay the brilliant clarity of a tropic sea. Below that a violet outcrop of coral heads. The water was seven fathoms deep.

I looked like some Mesozoic animal, for I wore blue flippered feet, a cyclops mask, and a silver metal cylinder. Two snake-like hoses led to my mouth. My smile became wan, for the fear was growing. I knew it was time to jump, but I hesitated.

Old terrestrial learnings urged resistance. To vault into the sea was to separate from air and to fill lungs with water. But I could delay no longer; it was time.

I leapt. Off into space. The white wave motion of the surface fell away. I was swept into warm embrace.

Such quietness. Such clarity. I was entranced. I lay there breathing slowly, narcotized by the sights around me. It was as if some pagan artist had splashed molten pigments across the sea floor. Blue striped grunts arched over small scarlet squirrel fish. Grey angelfish swam up for inspection and sighed away under a red coral forest. Purple gorgonians waved in the tidal ebb and beckoned me downward to chalk white sand.

For the first time I discovered the freedom of flying underwater. An indelible mental imprint. Suddenly I had free, weightless access to all three dimensions. I soared, swooped, and landed gently on a soft sand slope. I was one with fish and in love with my new liberty. I indulged all my senses as I had never before on land. An exhilaration crept into my cells. A euphoria never forgotten. The day of my consummation.

No bliss is without peril. My first dive brought an agenda of questions demanding resolution. How soon could I return to this wild new continent? Was there a career hidden somewhere in these blue-green canyons? How could I learn more? During the months which followed I released my yearnings in a flood of study about the sea and diving.

Two years later I rejoined the ocean just south of my first dive site. The Gulf Stream became my first classroom of study. In its blue, sunless halls I learned the basic elements of safe diving. I gained infinite respect for the unpredictable and crushing power of the sea.

One of my first teachers was a wiry charter boat skipper by the name of Captain Bob. He owned a slow-moving cabin cruiser which, for some incredible reason, was known as *Hot-Spur*. Each day he would take eager divers out to the reefs south of Miami. On hot, sunny mornings six or seven of us would gather at the old concrete pier where *Hot-Spur* was comfortably lashed.

Then began an essential ritual. An endless cortège of diving gear was lifted, carried, and passed to the stern of the boat. Tanks, diving-bags, masks, fins and regulators came and went in infinite procession.

9

This aspect of diving never changes, for heavy equipment must always be hauled into the boat, on your back, into the water, back into the boat, and then returned to the shore. It is little wonder that divers are strong; a major activity is this repeated weight lifting.

Hot-Spur rarely left the dock on time, for divers are notoriously late. Bob's passengers were no exception. However, by 9:30 or so, the last of us had stumbled on board, usually clutching a missing swim fin in one hand and a bag of soggy sandwiches in the other.

The trips from dockside to dive site were always amusing. Bob stood like a bronzed elf at the wheel of *Hot-Spur*. The boat seemed to steer itself, so often had it taken this morning journey past ships, docks, and sandbars. All Bob's ribaldries were spliced with his southern drawl and private chuckle. Soon he had everyone laughing and digging for the latest corrupt anecdote.

This initial pursuit of humour was necessary. A hidden menace of big waves sometimes waited for us beyond the last seawall. One could tell if it was going to be a serious ambush soon enough. As we approached the end of the channel, *Hot-Spur* would start an incredible rhythm. It was the drunken lurch of a sea-crazed animal. Its arrival splashed silence across the boat. Heads glanced down and grins appeared like puckered flowers. If the sea was really rough, a diver would sneak down below decks looking for a place to lie down. If he'd been out drinking the night before, he didn't stand a chance.

I, of course, was immune to all this. By skipping breakfast and secretly slipping Dramamine pills into my mouth, I maintained an almost normal pale yellow colour. Bob, born on a slippery deck, attempted to soothe everyone; in his quiet voice he reminded us that it was only an hour to the dive site. Once there, everything would be okay. In a boatful of churning stomachs, no one believed him.

But he was right. When we finally staggered over the side and into the supporting arms of the sea, a dramatic change took place. The sudden synchronization with the ocean's natural rhythms was like stepping on a solid shore. In a few minutes we all felt normal and stopped cursing Bob, his damn boat, and the sea.

Years passed. My ocean encounters increased in number and intensity. I passed from high school into medical school—barely.

Wherever I went I felt the call of the lonely sea. I loved its open wildness and the song that it generated deep inside. I love it even more today, sitting at the sea edge, looking out at the night waters.

The depths have spoken their unusual chorus to many others as well. In this same Mediterranean Sea, Edwin Link began a daring experiment in which a young Belgian diver lived and worked from a diving bell suspended for twenty-four hours, 200 feet beneath the surface. A few weeks later Jacques Cousteau completed another bold venture: for seven days two men lived and worked from a small station under 30 feet of water.

Thus, in 1962 off the south coast of France, man opened the gates of an age-old dream—to live within the sea as a free-swimming diver. These history-making dives were carried out not far from where I sit and muse.

In 1963 I had the good fortune to join the small band of men attempting to solve the medical and engineering problems of man beneath the sea. I went to work for Edwin Link. Although I could not fully appreciate it, I had quietly traversed one of the most critical crossroads of my life. My future with the ocean was assured.

For the next decade I lived and breathed the sea. I was as much fascinated by its beauty and hostility as by the systems and techniques we developed to overcome the challenges beneath its surface. I travelled to almost all of the seven seas and explored places of which my imagination had given no warning. I befriended a special group of individuals who were "brothers under the sea." In the process I fell in love with a mistress that for me breathed the pulse of the planet.

For those of us working in undersea technology, it was a decade of opportunity. These were years of great discoveries and engineering achievements. Because they were new to us, and to the world, they were cloaked in the exciting shroud of classical adventure.

I have been extremely fortunate. Over the years, I have participated in over a hundred underwater projects and expeditions. Many of them took place in remote and hostile places. All of them demanded the very best from men and equipment. In each operation I was an active participant and passive observer. It was the passive observer who stood slightly back from the events, knowing that he was watching the evolution of a new breed of human being—an individual suddenly astride a new techonology—underwater man.

It was always a team effort. Each man's life was suspended on the skill and experience of his neighbour. You will meet some of these men in the pages that follow. Indeed, it is to them that the title of this book most applies. Each in his own way *is* underwater man—a mosiac individual, unlike most terrestrial beings and with more than his share of determination and courage.

I knew as I worked alongside these men that I was sharing a unique human experience. I also knew that I was part of a story that would have to be told. It was the story of men working on the edges of a new frontier.

I have chosen eleven of my most unusual adventures for this book. I selected them to allow a glimpse into various kinds of underwater work and research currently underway in different parts of the world. Each adventure is contained in its own envelope of hazards—some similar, some different—but unlike anything found on dry land.

I stand up and begin to walk along the beach. On the midnight horizon a ship sparkles like a tiny diamond against coal. Its light is soft and comforting, for it clearly marks the present position of a lone traveller. It reminds me of my own life-journey, for the ship seems to hesitate as it moves cautiously into the farther blackness.

Writing this book has been an important pause for me. It forced the ordering of certain thoughts about the implications of man's recent undersea odyssey. It is an odyssey just begun. Ahead lie years of work and adventure. Man will move slowly but surely into deeper and polar seas. No one knows how deep he will go or how long he will stay. Such events are limited only by imagination.

Tomorrow is burdened with responsibility. The sea is still the most unchanged wildness on our planet. Underwater man must lead the way in its preservation and management.

Challenge is one of the essential nutrients of human growth. Many years ago I discovered that exploring the undersea world and its dominions is a splendid summons of physical and mental energies. Living safely within the ocean's harsh physical and chemical laws demands exquisite harmony between mind and body.

Fortunately, my own pursuit of the sea's challenges was supported by many people. Several were especially encouraging and influential. Their support led directly to this book, which would not have been written without their help:

Edwin Link, who conducted a young man personally below the sea and into his confidence;

Christian Lambertsen, Heinz Schreiner, Bill Hamilton, and George Bond, who worked so hard in the making of an undersea scientist;

Richard Birch and Michael Pitfield, who opened new avenues of thought;

Jack McClelland and Farley Mowat, who encouraged me to write;

Jennifer Glossop, my unsparing editor;

And Sharon McClure, my devoted secretary, who worked so hard to pull this manuscript together.

Throughout the writing of this book my thoughts were always close to my children, Tracy and Jeff. Their youth and love is continual inspiration.

It is impossible to name all those to whom I have become indebted during the past decade. There are too many. But I'd like them to know that I thought of them.

The sand is soft and cold beneath my feet. I stop and hear the sea begin to breathe. A low swell breaks darkly against the shore and then sighs away in recession. Another lifts its longing form up the sand. It retreats. The cadence has started afresh.

The lovely Aegean Sea, stretching its infinite night limbs in front of me, has moved to these and other rhythms for countless centuries. It has silently witnessed man's seaward march from beyond prehistoric times. It has felt the oar-pull of sacred triremes between islands, and seen the sails of Ulysses under its cliffs.

The Aegean's busy surface held together the people who lived here over 2,000 years ago. Its special quality played an important role in the forging of rational thought. Only yesterday, not far from this beach, I dove to a shipwreck and saw rows of ancient amphoria. They have lain on the seafloor for over a thousand years.

The sea is both past and future. It is an opportunity for man to seize the highest qualities within himself. It is a chance to find harmony with the major portion of the planet. The depth dimension is open, but much remains to be done.

The sky now softens the darkness above me with the earliest pale of dawn. In a few hours lovely white heralds of the sun will turn the sea into sapphire and invite all eyes into the hidden depths. Let us begin the descent. . . .

Mykonos, Greece, 1973.

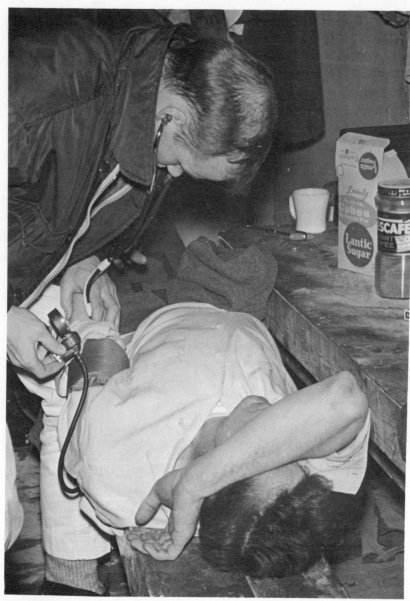

The author, examining
John McGean, who
is in severe pain from
decompression sickness.

Descent One
Internship
1963

The experience at first seems unrelated to the sea. Yet, in retrospect, the events of that bleak winter night in Toronto did much to direct me into the essential post-graduate studies that would allow me to become a diving physician.

Like many adventures, this one began innocently. I was sitting in the coffee shop of the Toronto General Hospital late at night. The day had been a long one. I was reviewing its major events.

Since July I had performed most of my work as a semi-automatic robot. Even when thinking became an option, I was so tired from being up all night that my muscles and brain moved only in their survival modes. A napkin calculation confirmed my salary at a little more than fifty cents an hour. I was, in fact, a lowly labourer in one of the most effective slave markets in the world. I was a junior intern.

Gazing into the future revealed a huge professional pyramid towering ahead of me. To climb to the top meant ascending a maze of steps that would stupefy the designers of Knossos. All was laid out, on a pre-ordained time chart calculated to burn out all my energies for the next twenty years. I could see myself turning slowly into charcoal on the altar of careerism.

For the past few weeks I had been surrounded by the delightful servants of the surgical service. Today had been a rotogravure of bleeding ears, broken noses, and rose-coloured throats. It was February, and cold enough for winter gusts to slide long arms under the cafeteria door. The sea was an infinite distance away.

My reveries were broken by the clatter of dishes dumped like iron ballast into a sink. I was alone in the coffee shop except for two senior

15

doctors who had just come in from their shift. They were discussing a case and occasional phrases drifted like pale leaves over the faded wood panel of their booth.

"A problem of recompression . . . the tunnel on Bloor Street . . . poor guy's first day. . . ."

I leaned over and carefully interrupted their conversation.

"Sorry . . . I heard you mention something about a pressure problem in a tunnel. Does someone have the bends?"

The elder physician answered easily.

"Sounds like it. But for some reason they seem to be having a lot of trouble with this particular case. With all the pressure they need to hold back the wall in those tunnels, there is always a risk that one of the workers will not tolerate decompression. If you're interested, they can probably tell you more about it in Emergency."

The white marble tile of the empty hallway echoed the soft hiss of my footsteps. I moved quickly, enjoying the freedom of motion that comes just before running. Suddenly I wasn't tired any more.

The Emergency Department was suspended in one of its rare periods of calm. Two young nurses bent over their cluttered station desk with telephones cupped close to their heads. Their voices were effortless.

"Yes, the man can be moved . . . he's on the surface . . . yes, we are looking for a pressure chamber. I know it's after midnight . . . but surely. . . ."

Two phones found their cradles almost simultaneously. A softly whispered "damn" flowed from under one cascade of hair.

The older woman looked at me with kindly eyes that had fixed on many a fumbling intern. An easy smile crossed her face.

"Now, what can I do for you, young man? It better be important because we're busier here than bears over honey. Talk up, or we'll have you examined for a speech impediment."

She was one of my favourites and ran this part of the hospital with a soft steel hand and plenty of humour. She knew more than most interns will ever know about medicine and people, and taught it to us gently. When she laughed, it was like a distant mineshaft collapsing.

"I heard that there may be a case of the bends, and that you're having problems with it."

"You heard right. We've been on these phones for thirty minutes trying to see if there's a pressure chamber available in Toronto. Both metro and provincial police can't seem to help. We've even contacted the Armed Forces. No luck. Got any ideas?"

My heart skipped a long beat. Tucked inside my wallet was a short blue card given to me some months ago. While looking into career

16

opportunities in diving medicine, I had visited Dr. Ed Lanphier at the University of Buffalo, one of the world's authorities in the field.

At the end of my visit, Dr. Lanphier had leaned across the table and said, "Here, take this, it's our diving emergency card. If you hear of a diving accident, use the telephone number on the other side. Our university decompression chamber is always ready, and so are we."

The card suddenly seemed to take on extra weight. Treatment for the bends is the same, no matter what its origin—sport diving casualty, or tunnel accident. In each case some elements of the blood stream alter their characteristics. Red cells tend to clump together and reduce effective flow of the stream. A major villain is small gas bubbles which can break into showers and block small arteries. The cause of bends is complex, and even today not well understood, but the effect on the body is devastatingly clear. A constellation of symptoms can cause laboured breathing, burning pain, and shifts of internal fluid. In its most severe form, bends can result in cardiac collapse and death.

It was 1:15 a.m., and getting late for whoever was suffering. I thought I would go out to the site to confirm just how bad things actually were. Perhaps I could do something to help. I called the senior resident on duty, Dr. John Kendal, and told him the situation.

"By all means, go," he said. "A little bold, but a good idea. Be careful with your diagnosis and treatment. Remember, you will be outside the hospital's jurisdictional boundaries."

A sobering thought. As an intern, I was not yet fully licensed to practise medicine. An error in judgement could turn into a lifetime albatross.

I shrugged a heavy parka over my intern's whites. Outside in the parking lot the wind placed its arctic arms around my shoulders. At first the car's engine spoke reluctantly, but finally agreed to turn over. The heater blew frost across my legs.

I drove down Toronto's deserted streets watching the wind sway the rows of old elms. After a few minutes I turned into a long twilight road just south of Bloor Street. The distance held a row of cars parked next to a cluster of slowly flashing yellow lights. As I drove northwards, the low silhouettes of dark sheds appeared to my left under the black skeleton of a small crane.

I parked next to the golden glare of what seemed like two large blinking eyes and headed for the only shed that had any warmth in the windows.

As I opened the door, I felt a wave of coffee-scented heat against my face. An old kerosene stove lay hot against the far wall and four naked bulbs burned along the plywood ceiling. The room held long tables and benches, and three limp coveralls hung on one wall.

17

About ten men, clustered as in a football huddle, stood over to one side of the room. Almost all wore parkas quilted with dried mud. A tired face turned toward whoever had let winter in through the door.

"What in hell do you want?" His eyes dropped from my face to my firmly pressed white pants. "You a doctor? I'm sorry. It's been a helluva night."

The huddle parted silently. I walked calmly up to the opening. A piece of ice slid into my heart. Lying on a bench and looking back at me, from eyes that had no focus, was the mask of pain. Creases, carved by agony's knife, crossed the man's brow and radiated into his lower jaw. His mouth was open, but no words came. I heard only the anguish of his soft and forced exhalations.

He moved slowly as if on some invisible rack. Arms and legs straightened and then his body doubled as if to condense the pain into something tolerable.

As I looked down at him, I felt the finger of ice move from my heart into my lungs. I started by examining him. It would tell me how quickly he was deteriorating, and would give me time to think. I reached for his wrist and found a pulse both rapid and weak. He winced as the cold finger of my stethoscope touched his skin. His breathing was shallow and laboured. I worked quickly and at the same time listened to someone beside me describe what had happened during the day.

His name was John McGean. He had reported early for work that morning, because it was his first day. Before going down to the job he was told about the procedures for getting into and out of the tunnel. First, there was the "transfer" chamber that would allow him to get into the shaft. He would climb down a steel ladder into this chamber, its hatch would be sealed, and the pressure would be increased until it was equal to that in the tunnel. Then he would walk through another hatch into the adjacent tunnel. After his shift, the sequence would be reversed. He was told that on the way back to the surface the pressure in the transfer chamber would be bled off slowly, according to a timed schedule. Like a diver coming back from depth, McGean would have to decompress.

McGean went to work by descending fifty feet down a steel ladder where he was "locked" into a large chamber. He heard a noisy inrush as the air pressure around him was increased to about twenty-four pounds per square inch. Then another large hatchway swung open and McGean stepped across into a long well-lighted shaft.

He worked for the next four hours shovelling wet earth into a bucket that seemed to have an endless capacity. He and the four other men strained to keep up the pace. It was hard and dirty work.

Just after mid-morning one of the walls slumped and began to cave in. Extra air pressure was pumped in to hold back the slurry of earth. It held. The men worked furiously to get rid of the unwanted mud.

About noon, McGean returned to the surface. He was tired but ate his lunch and talked with his new friends. Later, he was examined by a doctor for the routine medical required for all new employees. At one o'clock he returned to the dampness of the shaft and found that another cave-in had occurred. This one was really serious. Even more air pressure was required to hold back the loose-flowing earth. Eight and a half hours were needed to repair the damage. It had been a trying first day for McGean. He and his shovel had worked furiously to get rid of the earth that kept trying to fill the tunnel.

At 9:30 in the evening he was decompressed to the surface. He changed clothes and felt fine, except for an overwhelming fatigue. A few minutes later, walking to his car, he felt an intense pain begin in his upper leg and shoot like fire into his feet.

McGean hobbled back to the changing shed to tell the men. "It's as though someone is holding a hot iron on both my arms and legs. I have no strength. I stagger because my whole body feels weak and useless. When I sit down the pain is so bad that it makes me double up."

His partners moved quickly and got McGean back down into the transfer chamber where they recompressed him. He was held under pressure for about thirty minutes and then returned slowly to the surface. This treatment seemed to cure him so he drove home and took a hot shower.

Suddenly he began to tremble and shake. The pain returned, but this time in bone-tearing waves. A telephone call brought the police, who helped him into their cruiser and went looking for assistance. In desperation they located the doctor who had examined McGean earlier in the day. He recommended that he be immediately returned to the site.

At midnight McGean was back under pressure, but by one o'clock it was apparent that the second treatment was ineffective.

A black squall of pain swept over McGean. He doubled up and almost fell off the bench. Hands reached out to steady him. A long moan shivered over his pale lips. I was convinced. McGean was suffering an agonizing and possibly lethal case of decompression sickness.

I raced for the nearest phone. It was outside and about a hundred paces south of the shed. I secured the door to hold back the wind and dropped a dime into the cold metal slot. I asked for the number in Buffalo. Voices came and went and then I heard the phone ring. It sounded very far away.

A blurred voice answered slowly. It was Dr. Lanphier, wondering who the caller was and if he had any conception of the time. I spoke hesitantly.

"Dr. Lanphier, it's Joe MacInnis calling, and I wonder . . ."

"Joe who?"

"MacInnis. You may remember that I visited you some months ago at the university to talk about. . . ."

"Oh yes, I'm sorry. My thoughts are not together just yet. What can I do for you, Joe?"

His voice now had a sudden clarity to it. He recognized the concern in my voice.

For the next few minutes I detailed the night's events and finished with a description of how McGean last looked. The phone felt like cold steel against my ear. I began to shiver. I had run coatless from the shed and was starting to respond to the weight of cold pressing through the glass. Outside, wind blowing through the branches made the street lights tremble.

Dr. Lanphier's voice was reassuring. "Well, it sure sounds as if you've got a problem. I think you'd better bring McGean over here to our treatment facility. The pressure chamber is right next to a hospital and he can have the best of both worlds. I'll alert the team and we'll be ready to go by the time you get here. Better hurry."

I hung up the phone and raced back to the shed. Although bitterly cold, I felt much better. Buffalo had one of the best treatment centres in the world. At last McGean had a place to go. The difficulty would be to get him there in time. Buffalo was a hundred miles away in another country, and pre-dawn hours are never the best for rapid action.

I ran back to the shed and opened the door to a low keening sound in the far corner. McGean looked even paler than when I last saw him. A voice near the window rang out.

"A fire department ambulance just arrived, Doc. What shall we tell them?"

"Have them bring in a stretcher. We'll take him to Toronto General on the way to Buffalo. At the General we can give him something to ease his pain."

The stretcher and two men burst into the shed. McGean was eased between dark blankets and carefully carried out into the night. He grimaced as he was tilted down the steps, and felt the cold slap of the night wind.

I climbed into the ambulance beside McGean. The rear door closed securely behind us. The overhead flasher started to spin out its silent red signal. We picked up speed.

The ride through the night was a blurred swish of empty streets and occasional lone figures on grey sidewalks. A few glazed cars nosed their way slowly along the dim roads. Outside the ambulance I could hear the shrill wind. Tree branches waved at us like rigid tentacles.

McGean's eyes were blurry. He tried to sit up, but fell back with the effort. Pain kept him in continuous motion under the blankets. I leaned over and told him who I was and what we were trying to do. His whispered "thanks" only increased my feeling of impotence at being unable to relieve his pain. I reached out and held his hand.

We turned a sharp corner and stopped under the large EMERGENCY sign of the hospital. I saw it as an imaginary theatre marquee that only hinted at its real meaning. Behind that sign was a microcosm of the world in distress. The only thing more impressive than the accidents which arrived here was the skilful courage of the humans who repaired them.

All became motion, as the back door swung open and eager hands rolled the stretcher from its rack. McGean disappeared down a yellow hallway and I sprinted into another to search for a phone. In a few seconds I was again talking to Dr. John Kendal. He was deeply concerned about the events that I'd set in motion.

"Are you sure of your diagnosis? It's a difficult one to make and if you're wrong . . ."

I knew I was now locked into a serious situation. Decompression sickness can have many manifestations. There was also the possibility that McGean's pain was due to something other than improper decompression. My interior ice shifted to a new position. Although I had read a lot about "the bends," this was my first actual case. The ground under my decisions suddenly began to feel slippery.

However, I decided to hold fast to my convictions. All signs pointed to decompression sickness; and even if McGean's pain was not related to pressure, his return to the surface had been much too rapid. Full treatment had to include some form of pressure therapy.

"Yes, I am sure of the diagnosis. This man needs treatment in a pressure chamber, and needs it now."

In two minutes Dr. Kendal and I were standing together at McGean's bedside. I watched in admiration as the senior resident's eyes and hands moved quickly and positively over McGean. At the same time his mind ran down some hidden track where he sorted out the diagnostic possibilities. Here was years of experience in flow—adding, eliminating, and confirming. Kendal turned to me without a smile.

"I think you're right, but the real proof will be how he responds to pressure. I'd suggest you give him 75 mg. of Demerol to help his pain. Then you'd better get going."

While I administered the medication I heard Dr. Kendal speak firmly into a nearby phone.

"Yes, a fast ambulance . . . to Buffalo. Better get some help from the provincial police. . . . and alert the border authorities."

In another ten minutes I was walking out towards a gleaming chrome and white ambulance. Its engine was running and its red flashing light flinging out a powerful beam. McGean was lifted in through the open rear doors. I settled in beside him. The driver turned and motioned me to secure myself. His authority indicated that it was going to be a rapid trip.

We entered the open expanse of the street, and a city police car swung out in front. The ambulance accelerated down University Avenue and the lights of the side streets began to blur. As we raced through intersections I recognized corners that normally were filled with people and cars. As we moved swiftly in the slipstream confines of the police car, I could hear the dim wail of the siren. A few white faces paused to turn and study our rapid progress.

McGean was resting now. His breathing was slower and deeper and his body had ceased its uncontrolled writhing. He seemed almost asleep, except his eyes were rigid against the ceiling. He turned them to me.

"Doc, can you tell me where we're going?"

"John," I said, "We're going as fast as possible to Buffalo. It's the nearest medical treatment centre for your problem. We should be there in a little while. How do you feel?"

"Better, thanks, Doc. My body's getting kind of numb, but the pain's easing off now. Except in my legs. They hurt like hell."

We were a small caravan now. The city police car had been replaced by two Royal Canadian Mounted Police cruisers. They flanked us in front and behind. I suddenly realized that I was dead in the centre of three red flashers screaming down the highway. Somewhere over the horizon a team was scrambling. The ice shifted again.

I looked over at the speedometer. The needle showed ninety.

Bridges, trees, and houses were now blurred into a necklace of speed that lasted about an hour. McGean rested.

At Fort Erie the Niagara River begins to narrow and pick up speed for its plunge over the falls. I could see its black-mica muscles as we sped up and over the bridge which joins Canada and the United States. Our tires sung on the frosted pavement. McGean lay with his eyes closed, breathing easily.

We slowed to sixty miles per hour. And then to forty. A single uniformed figure stood at the entrance to U.S. customs. He waved us on. We squeezed through the cement columns and picked up speed.

In the strange city's streets we picked up a new escort. Two police cruisers from the city of Buffalo now paced us fore and aft like a mobile vise. All we had to do was follow. Every major intersection held a black and white car with its door open and a uniformed policeman holding back traffic. Cars were scarce at this hour, but several squatted in suspense to watch the drama pass.

I could see the long black horizon of the university campus ahead. We were almost there. I leaned over and spoke softly to McGean.

"How do you feel, John? Relief is just around the corner."

"Doc, I can't believe it. I feel just fine. I think I'm okay now."

A new finger of ice began to snake its way up my spine. A thought flickered like dirty smoke across my mind. What if, after all these miles and effort, McGean is already cured? What if he never . . .

We suddenly stopped. The ambulance was again surrounded by lights and motion. The rear doors were flung open, and McGean was lifted out and along a line of waiting hands. He then disappeared through an open basement window, coming to rest beside a large metal cylinder. I saw him reappear in front of a group of anxious doctors. Heading the examination was Dr. Bill Gillan, an expert neurologist, and an authority on decompression sickness.

It didn't take long to reach a decision. McGean's case was most serious and would require the maximum treatment. He was lifted gently off his cot and into the chamber. Dr. Lanphier stepped in with him. He had volunteered to help McGean through the long hours ahead.

During the examination McGean's pain had suddenly returned. The support team moved quickly. A round grey hatch was closed and its heavy metal latches secured. Dr. Lanphier confirmed that all was ready inside. A large valve marked COMPRESSION was turned and air hissed harshly into the main lock of the chamber. Through a viewport I could see both men holding their noses and attempting to equalize their ears to the new pressures. Thirty, forty, fifty . . . eighty, ninety and then one hundred feet.

The valve was closed and compression stopped. We could all hear McGean's voice over the intercom. It was faltering and his words sounded different in pitch.

"Doc, some parts of me feel worse. I don't know. . . ."

A hurried conference. Words exchanged with Dr. Lanphier. Decision. Compress according to the schedule down to 165 feet. The valve was opened again. Each viewport of the chamber now had at least two pairs of eyes straining through the plexiglass.

One hundred and sixty-five feet. Compression stopped. This was maximum depth. Sweat from the heat of pressurization rolled off Dr. Lanphier's forehead. The only sound in the room was the tick of a large clock.

Then McGean began to giggle. His low chuckle spread itself like oil into outright laughter. After a few words both he and Dr. Lanphier began to laugh uproariously.

Nitrogen narcosis. I was looking at my first picture of two grown men enjoying rapture of the deep. They were "high" on the increased concentration of gaseous nitrogen. Their laughter circled the room and splashed momentary grins across tired faces.

Quickly the laughter subsided. More words flowed between Lanphier and McGean. Then Lanphier's voice surged over the speaker. "I think we've got it. John says he feels much better. His pain has almost disappeared. I'll do another neurological examination and check things from that side."

In a few minutes the words came back.

"We're definitely winning. John's feeling a hundred per cent now, and his neurological exam is almost normal."

A collective sigh whispered to itself from each side of the chamber.

But John McGean was a long way from being completely cured. For the next forty-eight hours he would remain prisoner in his steel cocoon. Slowly, foot by foot, the pressure would be lifted off him until he reached the surface. During the ascent he would breathe different gas mixtures, according to the instructions given to him by Dr. Lanphier. For two days they would ride together up the long staircase of decreasing pressure. A friendship would grow. Like all alliances created under stress, this one would mature quickly.

I turned toward the well-lighted laboratory room with its banks of dials, analyzers, switches, and pipes. My feet moved slowly toward a cot lying behind a partition. I heard McGean's voice talking to Dr. Lanphier.

"I sure don't know how I can thank all you guys enough for all you've done."

"Forget it, John. We're really happy to help you out. And between you and me, some of those guys outside this chamber are secretly enjoying the excitement."

Descent Two
Treasure Island
1964

The sea lies mirror calm. Giant swells slide their blue bulk beneath the lazy bow of our ship. She is *Bluenose II*, on her maiden voyage, and now two months bound out of Nova Scotia. We are five degrees north of the equator and seeking an elusive dot of land somewhere to the west. Three hundred miles behind our wake is the jungle shore of Costa Rica. The Pacific lives its name. We are surrounded by tranquillity.

A snowy tern takes easy purchase in the sky. I see him from my cradle net, stretching full length under the bowsprit. Just below me the wooden bow slices water with a series of evanescent breaths. High above my perch two tall wood masts pierce the whispered wind.

Only one week ago, I had been slaving over a pressure experiment in Philadelphia. A post-graduate fellowship had enabled me to begin my studies of diving medicine at the University of Pennsylvania. It had been an exciting time, but the weather and work load had been such that when I received a call from the Canadian Broadcasting Corporation asking if I would join a filming expedition as an underwater photographer and doctor, I leapt at the opportunity. After all, it would only be a two-week interruption in my work, and the chance for adventure was irresistible.

On the horizon three smudges of rain cloud slowly appear. They release their soft grey colour and gradually become ominous. Beneath me heavier swells begin to lift the boat with sharper hiss.

From the central raincloud a flash of green appears and then melts away behind the lead sheets of a downpour. A few minutes later the cloud lifts a torn corner to reveal a soft emerald.

"Island! Two points off the starboard bow!" A voice sings from the rigging. Three figures cluster to the deck railing above me.

The island of Cocos,
with the *Bluenose* anchored
in the distance.

One of several early
maps of Cocos Island,
used to guide the
treasure-hunting party.

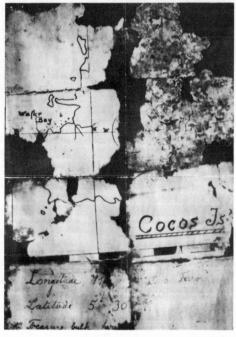

"There she is! It's her for sure! What a beauty!"

The island sits like the green crown of a submerged mountain. It is the only land in hundreds of miles of sea.

The first map of the island was sketched in 1541. The land was called Cocos because of the coconut trees that burdened its shoreline and interior. There is no record of the first voyager to discover this scrap of tropical jungle afloat in the Pacific. Probably it was a Spanish freebooter exploring the far water dominions of the mother country.

It is easy to understand why this oceanic speck was so elusive during the sixteenth century. Navigation was an art and not yet a science. Cocos was an island only three and a half miles across in a hidden sea. It had a certain ghostly quality that still remains. A ship travelling close by can be ushered away by strong local currents or blinded by heavy mists and rain-squalls. Many mariners passed within a few miles without glimpsing a trace of land.

The exact location of Cocos was at first difficult to pinpoint. Capricious sextants and geographers placed the island south, and then north, of the equator. Some ignored the ambulatory dot altogether, and said it didn't exist.

The first real visitors were probably whalers who left their steaming ships for temporary refreshment in the cool glades ashore. They sought the sweet water which splashed off the mountains and into the sea. Then they shouldered countless loads of wood, coconuts, and other fruit into their waiting jolly-boats.

Cocos might have continued its remote idyll indefinitely if a certain ship had not anchored under its shadow in 1820. The clatter and clank of shovels, spurred by the blood-greed of men, dug new meaning into her soil. Somewhere below her cool red earth, sweating and cursing men laid to rest millions of dollars of buried treasure.

Buried treasure! Two of the most exciting words in our language. Words which spark every dull eye, and quicken the pulse of the somnolent. An image heated with romance from the bright days of boyhood, when dreams are infallible. Many impassioned seekers continue to push back the curtains of skepticism. They nurture an illusive vision of locked trunk and gold and jewelled contents. Buried treasure! The thought of it stampedes men's minds.

As we approach, the island begins to assert its mysterious character. Green mountains tighten on a horizon blurred and streaked with rain. Cocos becomes a high, knuckled fist under the storm cloud. Then a golden arch begins to form over the northern coastline. It deepens with widening yellow, as the sun forces its heat through the raincloud. An indefinable red stain washes slowly into the edge of the golden arch. Incredibly a rainbow begins its ascent over the cliffs and into the vault of the sky.

I shift position in my hammock. There are no sounds from the rail above me. Separate thoughts dictate lone tunnels of silence. Some think of living colours in an unexplored sea. Others consider the wreck of a clipper ship waiting on the floor of a western bay. Most ponder the treasure.

The secret hoard of Cocos has haunted men's minds for over a century. It is a fable woven in myth and fact, murder and mystery.

In 1820 the wind-torn ship of a Portuguese pirate sailed into the western lee of Cocos. His name was Benito and he was anything but kindly. He was a pale-eyed, atavistic killer who pillaged the sea coast between Peru and Mexico. His was such a savage manner that even the most despicable of his crew detested him. For his own protection he spent most of the time locked in his cabin. He survived on deck only by carrying drawn cutlass and cocked pistol.

Immediately after anchoring his ship, Benito led a crew ashore to bury the combined booty of several rich hauls. Many large trunks were dragged across the beach and into the jungle. It is said the cache's value exceeded millions of dollars.

Some years later Peru flamed with revolution and civil war. The private wealth of Lima, and a splendid collection of church plate, was deposited for safety in the Fort at Callao. Then the fort itself was threatened. Over thirty million dollars in gold, silver, and jewels were transferred to an English sloop riding at anchor in the harbour. However, the temptation of such a huge treasure between decks was too great for captain and crew. They slit the throats of the Peruvian guards and sailed away on a moonless wind. The ship's name was *Mary Dear*.

The stealthy ones did not dare to seek harbour along the mainland and, by chance or intent, sailed to Cocos. They arrived on a course south of our own. The plan was to bury the bullion and wait. Jolly-boats were loaded, spades rang out, and the stained red soil accepted its second offering. Cocos became, without peer, the world's paramount treasure island.

Every bar in every sea port has at least one tale of treasure told each night, but the Cocos hoard has been the object of more words and dreams than all the others. However, somewhere in each man's telling must lie partial threads of truth, for the believers keep coming. The windless arrival of *Bluenose II* means that almost a hundred separate expeditions have joined these shores in the search for gold.

I climb atop the bowsprit for a better view. The sea and sky have suddenly come alive. A shoal of dolphins, those souls of lost sailors, appear in the water below me. In the sky are seabirds of every description; red-foots and boobies, frigate birds and terns. Some stay aloft while others seek our shrouds and rigging.

At last Cocos looms high above us, fresh and green from the recent wash of rain. The scent of jungle perfume carries across the water. The big schooner slows, and those hands on the forward deck watch for hidden coral. As we drift slowly in, all eyes fasten to the shore.

A small log raft lies on the beach, just up from the surf line. Four small figures appear from the jungle and begin to push it out to sea. Soon after they are water-borne, the men begin to shout and wave at the ship. Their voices are ugly and harsh across the water.

As the raft drifts against the ship, the four men crowd onto the forward logs and almost fall into the sea. They are a bearded, scruffy lot, ragged of cloth, and unbelievably anxious to see us. They cry out their greetings as they climb over the railing.

These men are the advance party of treasure hunters who have been on the island for several weeks. The most vocal is Ian McBean, leader of the group. He is living a twenty-five year obsession with the Cocos treasure. His voice rises as he reports his progress.

"I think we're onto it now. The jungle and heat have been bloody awful, but we're digging at three sites. The power digger has saved a lot of work, but it's so damn hard to haul it up the trails. The tangles of vines and grass tear hell out of your legs, but I'm certain we're on the track."

The man who organized our side of the expedition simply smiles. Maurice Taylor has come to make a film. He has all the ingredients of a classic—sailing ship, Pacific island, and buried treasure. He couldn't have typecast better. McBean is the ultimate treasure nut. He speaks of nothing else and he exudes confidence.

The party breaks off to go below decks into cooler shadows. We settle into the large central wardroom of the *Bluenose*. Glasses are brought out and the smoke grows. McBean commands the stage. He tells stories of maps, and Machiavellian motives. His outpourings are heightened by the rum flow, and the securely trapped audience.

One of the tales told during the long night involves a man supposed to have recovered part of the treasure. McBean is vibrant in the telling. His name was Keating, and he lived in St. John's, Newfoundland. In 1844 he took a strange and silent man into his home as a lodger.

One evening the lodger revealed a dark, fermenting secret to Keating, who had become his friend. He said he knew of a huge treasure hidden on a faraway Pacific island. He was certain of the exact location because he had helped in the burying.

McBean moves to the centre of the wardroom. His arms move like a conductor.

The stranger confirmed his story with a crude map which showed the island and its hidden hoard. Keating stared long and hard at the worn piece of paper. He had a good memory, but could not read or write. To Keating's dismay, the man later disappeared into St. John's foggy streets and was never seen again.

Keating then interested some local merchants in a project to find the treasure. They outfitted a small wooden schooner and set her canvas for distant Cocos. However, at sea bad blood began to flow. The voyage was filled with endless tempers, quarrelling, and despair. While crossing the Pacific the crew mutinied against Keating and the captain, a man called Bogue.

McBean has us now. Spellbound. He moves into a low crouch, hands still moving.

Lurching under arguments and spite, the ship finally reached Cocos. Anger flared deeper when the crew demanded a share in the treasure from Keating and Bogue. Pretending agreement, Keating and his scrap of paper finally made it ashore. He and Bogue slipped away one night when the crew was distracted by a sudden flow of rum and beer.

With Keating's map as a guide, the two men were able to locate part of the treasure and stagger it back to the beach. McBean's voice softens to a whisper. Then, suddenly the night wind heard the sound of a blow to the head, followed by one man's laboured breathing.

Silence. The story teller is caught in his own dramatic pause. He looks confused. The orchestra waits for the proper climax.

At this point the story dissolves like an old fading film. Keating somehow made his way back to St. John's carrying a few gold pieces and bars of bullion. There was a half-hearted attempt to try him for murder, but it failed for lack of a corpse. Keating died a marked and wretched man, surrounded by poverty and dreams of instant wealth.

Soon it became known that Keating had found treasure on the island. All the stories were confirmed; Cocos had treasure! The stampede was on. McBean had been one of the most fervent leaders of the stampede. He had spent a quarter of a century searching through libraries and talking to old-timers. The obsession dominated his life.

It is hard for us to share his single-mindedness—especially at that late hour. His audience slowly disperses, each person taking leave and retreating to his own thoughts.

The next day we make our first trip ashore. The jungle holds and hammers us with its heat. Our passage is always barred by vegetation through which we must cut and force our way. Four-fifths of the island is on sloping ground, and the trees stand at acute angles. All footing is sponge-soft with recent rain. Everything drips. After a quarter mile our spirits sink with the effort of simply moving ahead.

I watch the treasure hunters working to remove a huge palm tree lying in the way of their digging. Sweat streams from pale foreheads as they move the root-filled, stony soil from one pile to another. They remind me of a pride-broken prison gang working on a Georgia roadside. The air fills with vehemence and frustration. I withdraw to the seclusion of quieter jungle.

Cocos is an island of dense thickets and close-woven vines. Even in open glades, razor-edged grass lurks at knee level. Torrents of rain spill down the steep ravines, causing sudden landslides. The island has markedly changed its topsoil profile since anything was buried a century ago. Landmarks have disappeared or lie buried under tons of muck. In any event, a hand shovel dwindles into insignificance when pitted against the earthen tenacity of the island. If there is treasure in these rooty soils, the jungle owns it forever.

I find one of the rare passable trails. It is a narrow, rocky stream which leaps and swirls down from the high interior. I lie in it completely clothed and let the coolness trace and trickle over my skin.

A few feet farther down the hill the stream widens and I slosh through claret pools. Above are giant tree-ferns whose lacey foliage patterns the sky and holds back the sun. A flight of orange and black brassolid butterflies hover over a boulder sitting squarely in the stream. Both mossy banks are covered with hibiscus and clusia blossoms. I again lie on my back and listen to the water sing through the rootlets.

I descend to the last gurgling pool, where the stream spreads into the sea. A wandering tattler bird swings low out from the shore and gives a short cry. Below his free arc, *Bluenose* rides easily at anchor. Not far along the rocky beach I stumble onto a large boulder bearing a series of weather-worn inscriptions.

The earliest date I can read is 1797. Beside it is written: HIS BRIT MAG' SCHR LES DEUX-AMIS. This stone and others around it are mute sentinels to the travails and pleasures of men who have come to this island; lonely humans who thought they were in Eden or Eldorado. My eyes glance over some of the hand-carved letters.

Bk Virginia—Marks—A. Savvely 1875
Jos Grant Nantucket
G. Duffy Oct. 30 1843
J. Bond, Marblehead
Henry Hall of London
Fury
Mariposax 1:6 Px 1871 x 1870
Francis L. Steel Mar. 28. 1871
S. H. Harris   N. Bedford   Apr. 15-1842
Ship Alexdr. Coffin   D. Baker   Nantucket

Who were these men? What were they like? I will never know and can only imagine. The thought makes the brain weary with wonder.

My mind turns back toward the sea. I am eager to enter its cool folds. The first dive has special promise because we will be looking for the broken skeleton of a clipper ship, hidden somewhere between the hot sand and the anchored *Bluenose*.

The next day the Pacific closes its cool curtains over me. Gone are staccato voices and the roar of an earth-mad drilling machine. Gone is the stifling jungle and sauna heat. I am poised on a refreshing slide into new and silent depths.

Somewhere ahead is the shipwreck, still hidden by purple flues. She is a century-old tea clipper, broken by a storm. We know we are in the general area of her path, but uncertain if we'll find her.

The story of the clipper's sinking is a common one; the kind of tragedy which often befell those magnificent vessels that plied the tea trade from the Orient. She had been caught by a violent storm blown up from the south-west. The tempest had been so strong that it forced the big ship to retreat from the open ocean into this small bay. Her graceful lines had recently seen China, where she had picked up a load of tea. Her cargo was precious and her skipper a cautious man. He had thrown out two large storm anchors to hold the ship against a furious sea. Then the wind had swung around into the north and rammed huge breakers right into the mouth of the bay. The maëlstrom had been overwhelming. The anchors shifted and gave way. The ship was swiftly smashed by the high waves and talons of coral near the beach.

Maurice Taylor had suggested the bones of her hull would be found somewhere along the northern edge of the reef. This was my part of the expedition. Chris Chapman, the director of photography, and I had agreed that he would be responsible for the dry land photography, while I would lead the team of divers under the sea. Although my tasks were not central to the film, they were by far the more enjoyable.

As we enter the cool waters I think briefly of the others, up in the sauna heat of the jungle.

Thirty feet down, the seafloor swims into focus. I adjust the viewfinder of my motion picture camera. A tumbled pasture of flat mushroom coral surrounds large patches of white sand. As if on signal, Paul and I both stop. To the left and ahead of our swimway hangs a large tiger shark. He is motionless and in a deep sulk. His reaction to our intrusion is immediate. One convulsion of his harsh brown tail carries him into unseen waters. Two air regulators sigh simultaneously.

An unspecified curiosity impells us to the depression left in the sand by the shark's tail. On my approach I see flecks of rust mixed with granules of white sand. The two of us begin to scoop and fan with our hands. We are looking for the metal which originated the rust. Our motions are rapid. Ancient urges have taken over. Reward. My hand brushes against something solid. It is a round iron bar with a curved end. We tear at the sand. We have found the link of a huge anchor chain.

Further effort reveals the south-east lie of the long chain. We aim in this direction and begin a slow swim toward the island. We hope that the sand hides more links below us and that this general direction will lead us to the ship.

About eighty feet away, the chain emerges from the sand and climbs over a shelf of brain coral. Its age-worn links beckon up a gradual incline across a shallow coral staircase. Colourful reef fish sparkle the water as we disturb their claim to the shadows. A small green turtle scuttles to a new hideaway. We see little of this activity for we are hypnotized by the track of the chain.

I stop. The links have disappeared beneath a large coral outcrop. Its large size suggests a century's growth.

I become impatient with our slow progress and glance ahead into the pastel gloom. Ahead is a linear shadow. My blood quickens for there are few straight lines underwater, except those made by man. I touch my partner's shoulder. We lift off the bottom and swim steadily towards the unfocussed object.

A magnificent panorama unfolds ahead of us. The classic ship-wreck—broken into huge steel pieces of a puzzle, but seeming intact. Our eyes first grasp the splendid fractured arch of the clipper bow. It is broken, lying on its side and pointing ahead into deeper water. Its straight bowsprit was the linear shadow of my visual compass.

The main sections of the ship lie shallower, and are dramatically outlined by shattered squares of open cargo holds. A long row of curved ribs lies heeled over at an awkward angle. A helix of steel spars and plates tangles the sea floor. A great capstan tilts inside the round-ness of the stern. The ship reclines as if it had been rendered full length by an explosion. Parts of it are barely visible on the far water-sand horizon. Beyond is the crooked calcium edge of the reef.

Numb with excitement, we surface for a ten-minute planning session. It seems to take an hour.

Our second dive is a slow transit from bow to stern. For more than a hundred feet we marvel at a wreck that looks as if it were laid out by a Hollywood prop man.

As with most old wrecks, this one is a haven for marine life. Iron beams and plates are carpeted with coral and green-yellow algae.

Jagged holes provide security for moorish idols and butterfly fish. Sleepy groupers look out at us from larger wounds in the broken ship. And always the sharks. Ancient animals who remind us of prehistoric times and of a tomorrow when man will no longer swim the sea. They scull with millenial logic against currents silently sweeping in from the mouth of the bay.

These young "white tips" are plentiful in these waters. Although they are numerous, they are small and well-fed.

The next hour is spent alone in the euphoric solitude of exploration. We are joined by bubble-stream breathing and the whirr of cameras working. Occasionally a larger wave rubs steel against steel. It comes to us softly, a distant groaning sound.

We dive like men in a trance. Drifting across the twisted decks I try to picture the men who once worked here. I see them tall and brown and laughingly confident of their skills. I think of the last night. My mind carries broken images of the same men losing the fight against the tempest.

The ship contains guardians other than the white tips. A bilious moray eel is at home in the broken hollow that once was the bowsprit. Normally such eels do not bother us, but this one has a fire-plug head and an endless body. His flinty eyes speak malevolence, and I retreat with respect.

Just as we are about to leave the water, a cloud of milky sand drifts toward us from the shallow waters. Puzzled, I swim forward to seek its source. Slowly a giant manta ray takes shape in the fog. He is ten feet from wing-tip to wing-tip. He soars magnificently. His huge torso drifts trails of powdery substance as he moves forward. He has been lying on the seafloor and is slowly removing his sandy disguise.

The giant pauses and beckons with both black pinions. He stares at us through the two horn-like appendages that give him his devilfish name. Then he revolves on arched wing-tip, and takes slow flight.

An inner reflex sings out. I follow. I swim as fast and discreetly as possible, for I want him on film. It's no use. For the next five minutes, he occasionally ripples a wing in disdain of my attempt at speed. I kick my legs in fury at the heavy water, and heave on my regulator.

My breathing becomes asthmatic.

I finally squeeze off a few frames. They are certainly not close-ups. When viewing the films I will detect a sleepy, sardonic smile. I have chased an animal whose easy progress mocks my own.

Air exhausted, I rush to the surface. For the next minute each inhalation is embraced with greed. After the gasping, I swim slowly back to our small boat. My right of sea-trespass is as slender as ever.

The days at Cocos wheel by in a velvet procession. Soft winds, blue-black rain squalls, and silver and moonless nights. Time begins to flow through that perfumed sea-bliss found only on a remote Pacific isle. The crash and fall of breakers, the purl of wind in palm trees, all becomes a part of earth's vast silence. Time is counted not by watches or dates, but by events—the day of the huge rain storm, the visit of the hammerhead shark, or the finding of the old anchor.

For those of us on the dive team, the time is mostly spent under the sea. They are moments filled with the rush of joy and the pleasures of each discovery. Not so for the treasure hunters. Each day frustration mounts. Each night finds them more vocal and depressed about the lack of progress. Sweat drips. Time is running out.

Notes from my Cocos journal echo those days. They seem so close it is hard to believe they were written ten years ago.

February 15th
Today, while the shore party again fought the jungle and its heat, we explored the chill recesses of a place we called "the corridor." It is a narrow, sea-swept tunnel through a thin peninsula which reaches out to Cascara Island. It is almost 300 feet long and opens at both ends to the swirl of the sea. Its jagged volcanic ceiling is about thirty feet high, and in places its torn walls are only about ten feet apart. Its floor is the sweep and pull of the ocean. Everything is in black shadow, even at high noon. As we swam through, the sea undulated madly, like a long black serpent trying to avoid capture. Midway down the tunnel I wished I hadn't come. Twice I was almost smashed against the walls. The sharp shaft of sunlight waiting at the far end was the only thing buoying my spirit. Needless to say we didn't make the return trip.

February 16th
Today I had a brief attack of "treasure fever." Cruising the surface I looked down to see a "sea chest" half buried in the bottom sand. My pulse jumped. I swam down to a simple and empty box. End of treasure fever. Glad I didn't shout out and make a fool of myself.

The box wasn't really empty, for it held a long spotted moray eel. I filmed for a while until the landlord became agitated at my closeness.

The shore party came back on board tonight even more concerned about their lack of success. No wonder. By now they must have moved tons of red earth.

February 17th
Awoke today to an incredible downpour of rain, and the shrill of hundreds of sea-birds in refuge on our masts and rigging. Their soggy screaming and flapping is one of the most unusual animal activities I've

ever seen. During the downpour we stripped and soaped ourselves, laughing and singing like irreverent children. It drizzled rain all day and a cloud of red mud swept out of the river. We dared not dive, for the visibility had dropped to zero.

This afternoon I treated one of the men for a wild pig bite. Medical school didn't teach this one.

February 19th
The rains returned again, in unbelievable tropical fury. After the storm swung out to sea, Alex and I rowed over to a rocky, wave-smashed beach. He set me ashore and I clambered over huge boulders to the base of a new waterfall. This sudden cataract appeared during the rain, and was the work of water spilling from the jungle floor on top of a four-hundred-foot cliff. What a sight at the base! Spray swords, lashing the trees and bouncing off rocks. The wind created from the falling water is at least twenty miles an hour.

In the afternoon we plodded to a hot humid place that McBean calls the Graham Bennet diggings. We crawled into a series of fearfully dark caves. I don't know how the treasure hunters can work in such hot and confined quarters. Like working in a Singapore storm sewer.

February 20th
Little time left. Today we explored the waters at the base of Conic Island. The sunken cliffs seem to plunge down forever into the purple haze. We went down several hundred feet with our cameras and all three of us felt the warm arms of nitrogen narcosis. The water was unbelievably clear, and an endless school of blue tangs made a long route-march in front of us. These waters and animals are the clearest and most colourful I've ever seen. On the way back up, several "white tips" flashed between us. During the last few dives the sharks seemed more nervous and disturbed. Perhaps we should leave now, before anything happens.

February 21st
Our last day. A series of frantic efforts and scrambles for the treasure hunters. I feel sorry for them. They have spent many weeks amidst the beauties of this place—and have been too busy or tired to notice the wonders. All they know about Cocos is fatigue and sweat, and the chimeric treasure. Their fever blinds them. . . .

And so ended the days of tropic sea rhythm. At night we were rocked by the swell, and in the dawn came out on deck to the rolling heights of the island. *Bluenose* rode silent on the shoulders of an unchanging ocean. A sea wind continually blew. It was a breeze scented with salt and kelp and occasionally the rock-damp of air in a cave.

Cocos was an extravagance on the senses; even as we sailed away, her green towers paraded soft majesty over white wave-songs. It was an evening departure and the fading island held out low and wistful memories.

Within a week I was back in Washington. It was March and it was cold. Soon I was again engrossed in my work. At their Experimental Diving Unit, the U.S. Navy was conducting an extremely significant deep diving experiment. Two men were living at a pressure of 400 feet to ascertain the effects of such an exposure. It was the key study for work that was to change my life.

The deck decompression chamber being lowered into position aboard *Sea Diver*, during final preparations in Miami. This chamber would house Stenuit and Link during the 4-day decompression.

Descent Three
Preparations
1964

Edwin Link. A name synonymous with the romance of undersea exploration. Former aviation pioneer and businessman, he now devotes his life to inventing new devices to allow divers to work deep beneath the sea. A man of ceaseless energy for his passion. In recent years no one had done more to advance technology for underwater man.

I stand on a sun-washed pier in Key West. Behind me are the glittering crossroads of the open sea. In front of me is Edwin Link, walking with authority across the stern of his ship. Link makes the final adjustments to his diving gear and stands at the edge of the ship, which faces the harbour. He does not hesitate. In a motion that speaks determination, he leaps easily into the water.

I watch the bubbles released by the first breaths from his aqualung. They softly agitate the surface and then leave it undisturbed. Only quiet rings and small calms remain.

My gaze carries back to the stern of the ship. She is Link's own research vessel, *Sea Diver*, and the embodiment of all his dreams. Since 1959 she has carried Link to countless confrontations and adventures with the sea. Her one hundred feet of white gleaming length echo the same ocean confidence as her owner. She is Link's from the keel up. Every hatch and deck plate had been planned to support his undersea endeavours.

Behind the spot where Link has departed is an extraordinary array of diving equipment. It includes breathing systems, ballast trays, and a long aluminum chamber that sparkles wetness in the sun. The chamber is an underwater elevator, used to carry and protect two divers deep within the sea. This submersible decompression chamber, or SDC, as Link calls it, has just been lifted from the water. Warm silver rivulets run down its sides and into steaming pools on the deck. Overhead, a large black lifting boom suspends long cables swinging black in the sun.

No more bubbles come from the spot where Link has dived. The harbour's surface is now unbroken except for tiny cat's paws of wind. To the casual observer it is as if Link had vanished.

He has, in fact, disappeared into a place of his own creation. Lying some thirty feet below the ship is a small underwater station. Link, who has designed it, is now inside it and sharing its air with one of his crew. There are two small strange structures at the bottom of Key West harbour. Link is in the first one, called SPID. It is a short black cylinder with an entrance hole beneath it. It is made of rubber and about the same volume as a two-man tent. The other structure is slightly larger. It is named *Igloo*, because of its hemispherical shape. It too is a Link creation, although he has borrowed the name. If an observer could lift the water away from this section of the sea he would gasp in disbelief. It is the stuff of science fiction.

For several days Link and his team have been conducting a long dive in the shallow waters beside the ship. They are almost finished. It is part of a slow crescendo of effort, leading up to the world's deepest long dive. In the protected waters of the harbour the team is trying to simulate the problems that lie ahead. It is a chance to shallow-test all the underwater systems and techniques. After eliminating the inevitable difficulties, they will move to a deeper, but still intermediate, depth. Some months from now, after repeated tests, Link and his team hope to challenge the deep ocean with confidence.

I move quietly to the corner of a long wooden crate. The sun is warm on my shoulders. I slip my shirt off and sit down to wait. I am not expected on board *Sea Diver* for the next hour, and there is a certain delight in being able to sit back unnoticed and look at my future place of work. All is quiet on the after deck of the ship. A lone figure in the deep shade of an awning watches a small control console. I can hear the muted voices of the divers below coming in over the intercom. My mind drifts back to the events which have carried me like a rushing stream to Key West.

It began soon after I left the hospital and my junior internship. I knew that the United States was the only country where I could pursue my interest in deep-diving medicine. In addition, there was only one man who was doing the kind of work that really interested me. He was a dynamic inventor, who recently had been developing systems to allow divers to live and work for long periods deep beneath the sea. He was a living Jules Verne. He was Edwin Link.

But Link was as remote as the rings of Saturn. For months I had tried to contact him by mail and telegram, but his schedule always kept him just ahead of my reach. In addition to his own undersea development work, Link had just joined the review board investigating the sinking and loss of the nuclear submarine *Thresher*. In the fall of 1963 he was in Washington, moving from meeting to meeting. Each autumn day burdened him with more and more responsibility.

One morning I hesitantly picked up the phone and placed a person-to-person call to Link. After a three-hour search I was finally able to talk to him. I told him I wanted to come to Washington to inquire about joining his team. He agreed to a fifteen minute appointment the next day at ten o'clock.

I was exultant. After six months of trying, I would at last be able to tell Link personally how important it was for me to work for him. My mood did not last. I knew I would have to be convincing. Fifteen minutes is a short time.

We met in an old red-brick building at the Washington Navy Yard. From the window I could see the dark and icy currents of the Potomac River. In some of the nearby buildings, the United States Navy did much of its basic medical research into the effects of deep diving. Large and complex decompression chambers hummed with men testing themselves against the gas and pressure conditions found deep beneath the sea. Their findings represented the most advanced edge of manned diving technology.

As we sat down, Link fixed me with a stern eye. My shoulders sagged, as I suddenly felt the weight of confronting a man and his image.

I opened up like a verbal air hammer. In three minutes I had said everything. I told Link of my uneasy year of internship and my love for the sea. I emphasized repeatedly my concern for the bold men who wanted to live and work within it. I brushed by quickly any of my diving accomplishments. In this room, with this man, they seemed impressively pale.

Link just smiled and began to reveal some of the energy and enthusiasm he has for his work. He started to talk of his future plans.

"We'll stay here in Washington until the review board has finished its meetings. Then I'm going to take *Sea Diver* down to Key West. Sometime in late spring we hope to be ready to place two men at 400 feet on the continental shelf. We'll do it in warm clear water to minimize the difficulties of such a first step. Probably somewhere in the Bahamas. For the past few months we've been working on a small rubber structure that will support two men for several days. Its name, SPID, stands for submersible, portable, and inflatable dwelling. We'll use our aluminum submersible chamber to carry the two men to the bottom and back to the surface. There'll be a larger deck-mounted chamber for them to decompress after the dive. It'll be a long trip home. They'll need room and comfort during decompression. Can you imagine decompressing for four days?"

The whole scheme had an incredible ring to it. It seemed an improbable, even impossible, challenge. I felt as if I was looking through a submarine telescope into the twenty-first century.

"There's a lot of work to be done, especially in physiology, but we're getting a great deal of help from both the U.S. Navy and the University of Pennsylvania. Our biggest problem is in diving medicine and life-support. How'd you like to come on board as our full-time doctor?"

The question shook me like two rough hands. My willingness must have been evident. Link laughed and said, "Good, I am counting on that huge enthusiasm you seem to carry. You'll need it, for it's a tough job ahead. During the next few months you will work under Dr. Lambertsen at the University of Pennsylvania. He knows more about diving medicine than anyone in the world. When we're ready, you join us in Key West. Then we'll go and tackle the deep dive."

I left the Navy Yard exuberant beyond belief. I had just been accepted as an apprentice into the world's most distinguished fraternity.

The exhilaration sustained me during the dark winter months in Philadelphia. Dr. Lambertsen and his staff patiently guided me through the basic problems of undersea medicine. They showed me the hazards which make man vulnerable underwater, and the preventative steps needed to ensure his safety. They outlined the steps in preparing life-support systems and treating emergencies. It was one of the most rewarding and instructive periods of my life.

Dr. James Dixon became my personal mentor for a series of high-pressure studies on small animals. The experiments were set up to test gas analysis equipment and to investigate the effects of extremely high pressures on mice. The results would provide clues to larger animal and possibly human responses. I would also gain experience with the oxygen and carbon dioxide analyzers we would use on the deep open-sea dive.

My "workshop" was a short, squat barrel of a chamber bristling with a porcupine quilt of equipment. Lights, pipes, thermistors, gauges, and wires worked weird pathways in and around its shell. A bank of high pressure gases stood nearby.

The most exciting moment came late on a March morning when I looked into its interior to see four mice moving slowly about. The pressure gauge above the chamber read four thousand feet. Almost a ton of pressure on each square inch of white fur. Four small animals surviving pressures found almost two-thirds of a mile under the sea. They breathed a fraction of oxygen—about 100 times less than I was inhaling. The gas surrounding their small bodies was unbelievably dense. It must have felt like breathing thin soup.

The mice were not acting normally. They moved with faltering steps and squinted at me with half-closed eyes. However, the fact that they were able to survive at all suggested that man, too, might penetrate much greater depths. I knew we had a long way to go before man could step out of a diving bell at 2,000 feet, but here was the first modest indication of the kind of pressures man might be able to tolerate.

Mice are far less complicated than humans, and yet our experiments were not a simple task. We spent many long days investigating the limits of the tolerable mixtures of gas, the various internal temperatures, and the rates of compression. We had lost many mice when one of these factors was not exactly right. In order for men to reach this incredible depth, just the right life-support techniques had to be employed. It would be a long and difficult research journey.

Decompression would pose other problems. The stress effects of bringing animals back to the "surface" can be fatal. I suddenly realized how much more difficult it might be for men to survive, and I shuddered.

The warm Florida sun scatters my thoughts.

The water next to *Sea Diver* begins to boil into small pools of salt froth. A diver is about to surface or someone is venting the air space of the submerged structures. Two, and then three heads appear. It is Link and his divers. They are smiling.

I climb down to *Sea Diver*'s hot grey deck and wait under the awning. Link comes easily up the ladder and over into the shade. He sloshes a trail of warm water behind him.

"Joe, I'm glad you've arrived. I'd like you to meet some of *Sea Diver*'s crew." Several other young, tanned men appear from the ship's forward section.

"This is my son, Clayton, Robert Stenuit from Belgium, Dan Eden from England, and John Marguetis from Greece. We're proud of the fact that this is an international team. Your being from Canada makes it even more so. Welcome aboard."

I move slowly across the row of firm handshakes studying the young faces above them. I especially remember Stenuit: short, serious, the lead diver. Eden: tall, slender, the ship's engineer. Marguetis:

strong, stocky, the ship's cook. Clay Link, tall, blonde, resolute in stature. The whole crew has an unspoken air of purpose and camaraderie. I suddenly feel the vulnerability of the newcomer.

It is a sensation which does not last long. Within weeks I have earned the friendship of the small group of men. We grow to respect each other's skills and ambition. Our mission binds us together in the rare adhesive created when men confront a difficult objective.

We are facing some long and serious odds. There are hundreds of problems to be overcome before we can safely carry out the four-hundred-foot dive. Weeks, and then months, pass by. Each day is filled with minor successes and major delays. We often solve one problem only to find a larger one looming behind. Our preparations are more than simply "getting ready." We are, under Link's steady compass, developing new techniques and systems each step of the way.

Fortunately, the U.S. Navy experiments have already eliminated one of our most serious hurdles at their Experimental Diving Unit in Washington. Their simulated long-duration dive to four hundred feet parallelled the dive profile that Link planned for the open sea. Under controlled conditions, possible only in a laboratory, men and equipment were tested under the watchful eyes of over a dozen scientists.

After my return from Cocos, I worked at the Unit with Jim Dixon for almost a week. When I arrived in Key West, I reviewed the dive details with Edwin Link.

"Two Navy sailors climbed into one of the large chambers and were gradually pressurized to four hundred feet. They remained at this depth for twenty-four hours breathing four per cent oxygen in helium. While on the bottom they suited up and climbed down into a round pool to conduct swimming and other underwater work tasks. We saw them easily through the viewports. At all times they seemed perfectly normal. The only apparent difference from the surface was the high pitch and discord of their voices. They sounded like chipmunks in heat. I spoke to both the men after the dive. They hadn't noticed any ill effects. All post-dive tests confirmed this.

"Jim and I were primarily responsible for monitoring the oxygen and carbon dioxide levels. Jim set up the analyzers and we watched them every half-hour for almost a week. It was critical to ensure that the divers were maintained within a safe oxygen range.

"I was really impressed with the length of the decompression schedule. After twenty-four hours on the bottom the divers became 'saturated.' Their body tissues were almost completely 'full' of the gas that they were breathing. It took almost four days to decompress—an unbelievably long process."

44

Link listened quietly. He got up and moved across the floor. He was excited.

"Well, Joe, you saw your first deep saturation dive. I predict it's just the start. Over the next ten years there'll be hundreds. It's going to open up a new field of diving. Ours is a big responsibility. We'll be the first to confirm it can be done in the open sea."

Link is right. Military, commercial, and scientific teams will discover that the "saturation" technique allowed divers to remain at a given depth for an infinite period and then make one "final" decompression. One single ascent will replace the repeated decompressions necessary for short-duration dives. Saturation means increased time on the bottom, a savings in total decompression time, and increased diving safety. It is a major breakthrough.

Saturation was not a cure to the problem of decompression, but it did mean that the number of returns to the surface could be reduced. The danger of decompression results from the inert gas (nitrogen or helium) in the breathing mixture, which enters the body's tissues through the bloodstream. If the pressure is reduced too quickly, bubbles form in these tissues, in much the same way they do when a bottle of carbonated water is opened. Sudden decompression from a long, deep dive can be fatal; even a slight miscalculation of decompression requirements can cause serious injury to the joints or the central nervous system. A diver must therefore be decompressed slowly, according to a careful schedule, so that the inert gas can be washed out of the tissues by the blood and then exhaled by the lungs. Whereas the demands of decompression become more stringent with depth, with time they increase only up to a point. After about twenty-four hours at a given depth, the tissues become essentially saturated with inert gas at a pressure equivalent to the depth; they do not take up significantly more gas no matter how long the diver stays at that level. Therefore, if a diver must descend to a certain depth to accomplish a time-consuming underwater task, it is far more efficient for him to stay there than to return to the surface repeatedly, spending hours in decompression each time. Although this "saturation diving" is efficient, it imposes an extra technical burden, because the schedules for the ultimate decompression must be calculated and controlled with particular care.

I look out *Sea-Diver*'s window and into the sea. If only she will keep a calm, serene face for us. . . .

One of our remaining problems is to devise a practical method of removing the carbon dioxide which will be produced inside the SDC and SPID. A diver exhales about a cubic foot of this gas every hour. On the surface, carbon dioxide is quickly diluted by the surrounding air. But in close spaces, such as diving helmets and underwater stations, the gas builds up. Toxic levels are soon achieved and lead to breathing difficulties and unconsciousness. The gas must be removed with a carbon dioxide scrubber system. However, in 1964 there is no such system. We need something small and reliable. We need it quickly. Otherwise we cannot consider the deep dive.

Link's solution is typical. One day he comes on board with a small package. It is the electric motor of a standard vacuum cleaner. He places it on the work bench. It is the perfect motor. Its exhaust can be used to blow gas across the chemicals that remove carbon dioxide. We encase the whole device in a metal cylinder three feet long and eight inches in diameter. Suddenly we have a reliable scrubber. It will run for almost a full day before we need to replace the chemical. Quickly, the problem has vapourized.

Link insists on making the first test himself. He climbs into the submersible chamber and secures its aluminum hatch. He waits inside for over an hour. At this point his carbon dioxide has built up so that it is easily detected. He is beginning to feel breath hunger.

I am on deck monitoring the operation. I hear Link plug in the scrubber. Click. A faint whirr.

"Okay, topside. Beginning the test. The scrubber's on and. . . ."

Link's voice ceases abruptly. It is overwhelmed by a series of deep racking coughs. The scrubber stops. The coughs increase in intensity. Then they too stop. Silence. The sound of a body moving.

I race to the chamber and unlock the hatch. Link sits looking up at me, a snow fall powder almost completely covering his face. Tears run from his eyes. He grins sheepishly.

"After I turned on the scrubber I bent over to check the blower's effectiveness. A thin layer of fresh chemical blew right down my throat. I need some water. It tastes like old wood ashes."

He gets up quickly and clears the hatch. As he walks across the deck I hear him say to himself, "She sure throws a strong breeze."

Link tries again, and it works. He runs the scrubber under pressure and is successful. Three of these new systems are built and placed in our chambers. Two others are kept as spares.

Working on the after deck of *Sea Diver* opens my eyes to the enormous energy of the ocean world. It is tomorrow's battleground. A long rack of "k" cylinders contain breathing gas stored at over two thousand pounds per square inch. Tons of lead ballast squat in a corner. Huge lengths of rope and chain lie, coiled and free. Overhead swings the black muscled arm of the lifting boom. Under our keel waits the swift destruction of the kinetic sea.

After the shallow water trials are completed, Link decides to move his underwater structures up onto the pier. The heaviest object is *Igloo*. Its circular ballast tray holds over thirteen tons of lead.

One morning a large yellow crane rumbles down the pier. A prehensile steel animal. Its thick cable is lowered into position over the water spot where *Igloo* lies hidden. A diver drops to the bottom and shackles its free end to the centre lifting wing. He surfaces and swims clear.

The crane operator eases up the slack in the cable until he feels tension. *Igloo*'s back breaks the surface like a slow surfacing whale. Air gushes out of the broken surface of the sea. The glistening bulk lifts free of the water until the crane holds the full weight in the air. The crane operator then drops his boom slightly forward for better leverage. Six more feet to clear the dock. Up . . . up . . . slowly it comes. Thirteen tons of lead in a ringed tray beneath the black skirt of the structure. The yellow crane suddenly looks fragile. It seems to have captured a huge sea monster by a delicate thread.

The boom swings a short arc over the water. *Igloo* is over the pier. Then begins a gentle forward descent. But the slight action places too much weight on the top of the crane. Its rear wheels start off the ground, like a horse getting ready to kick. The crane operator feels the new motion. *Igloo* and the crane will topple into the water. He releases *Igloo*.

Tons of lead and rubber crash onto concrete. Air hisses from collapsing rubber walls. Heat waves shimmer.

A moan echoes across from *Sea Diver*. Whispered curses take explosive wing. Years of work lie shattered on the dock. The crane operator drops his head on his arms.

We walk slowly toward the torn rubber and fractured trays. Clay Link looks down wearily at the damage.

"A fine undersea station—but a lousy parachute."

Some weeks later *Sea Diver* departs Key West. *Igloo* is left behind. It is beyond repair. A major feature is eliminated from our deep dive program.

I quickly gain an enormous respect for the potential and kinetic energy associated with our work. I learn real reverence for high-pressure cylinders, which explode like bombs on impact. I learn to stay well away from suspended objects and to keep my unwary feet out of lines coiled on deck. A short length of nylon rope can be surprisingly lethal. If its free end is connected to an anchor and falls overboard, your body could quickly follow. Like a desperate python, the rope doesn't care whose leg it surrounds.

I soon discover that the sea does not release secrets easily. Just when she has seduced us into believing we have control she stuns us into submission. She is always unpredictable and at times displays the malevolence of a cornered tornado.

Our journey north to Miami is a brief, but peaceful change. Key West and the constant activity of the shipyard lies far behind. It is May, and the Gulf Stream carries us gently north. The soft airs of fresh trade winds blow across the deck. To the west lies the hazy necklace of the Florida Keys. We move forward on blue skies, blue water and the breath of summer.

Near Alligator Reef we stop to carry out a dive that will give us additional confidence in our equipment. We slip out of the main currents of the Gulf Stream and anchor close to a low green island. Four anchors are laid out from both the bow and stern, so that we hang securely in a four-point moor. The hot sun compresses the ocean into swimming-pool calm.

Our task is an easy one, and we work with confidence. Our objective is to lower SPID sixty feet to the ocean floor, secure it on the bottom, and then retrieve it. First we have to lift the deflated structure over the side and gently lower it into the sea. We will fill it with compressed air and ride it quietly on the surface. It will be a simple matter to load the lead ballast. The tray will be suspended only a few feet underwater. When enough ballast has been added, the station will slowly sink. We will control its descent with a thick nylon rope fed out along the ship's boom. After SPID is on the sea floor it will be an easy task to fill the ballast tray with more lead.

The first steps are effortless. The rubber tent floats like a bloated whale, rising gently with each soft swell. Her hot black carcass is over-laced with a net of thick nylon line. This secures the tent section to the ballast tray. As we add bars of lead, the structure slowly submerges. As a precaution the nylon line coming down from the boom is shackled into the central lifting ring. It lies like a white guardian snake over the slowly receding bulk of the tent.

I take my turn loading the lead. I swim to *Sea Diver*'s ladder, where someone hands me a twenty-five pound bar. Swimming is unnecessary. I drop through the blue water with an uncomfortable speed. With one hand I hold the lead. With the other I reach for the top of the tray to retard my fall. I make the rapid trip so many times that my ears begin to pinch. On my last dive I look at the heavy tray of lead and the swollen rubber skin. Eight tons of weight hang like a pendulum below a thin bubble of air.

I climb the ladder and start across the deck. The air jumps with a muffled explosion: the torn sound of escaping air. I turn to see an opening shred across the exposed rubber. SPID's skin has given way. For a few seconds air roars out of the hole. The wounded structure gasps a huge bubble and disappears.

I stand frozen to the deck. Then begins a chain reaction. The nylon line holding SPID snakes out to its full length, and comes to a complete and shuddering stop. The boom and the ship quiver with the impact. Eight tons suddenly in the sea. The ship lists slightly and the nylon line sings with the new strain.

No one moves a muscle. Eyes travel along the newly formed avenues of stress. The nylon line, thin with weight, runs down the boom and across the deck. It then makes a right-angled turn around a large deck plate pully. The shackle holding the pully is not designed for such sudden loads. A steel ring snaps and the shackle tears away. The nylon line is now a giant sling shot. Several seconds later the pulley lands with a torn splash a hundred yards off the stern.

Link's voice scalds across the deck. "Cut the damn line."

Dan Eden reaches for a knife. With such an enormous weight on the nylon he only needs a few strokes. The rope parts with the crackling of distant gunfire. The free end smokes its way over the pulleys, along the boom, and into the sea.

We dive down, to find SPID limp and defeated on the ocean floor. A jagged tear runs two feet along its broken spine. Piece by piece, we return the lead to *Sea Diver*'s deck. We then lift the station over the side and into its cradle. Everyone stops by to look at the gaping black wound. It had been started by the abrasion of a steel cable. Under continuous tension from the sea, the nearby cable had worn a fault into the rubber wall.

Nothing is said. Some silently wonder what would have happened if one of us had been working on the ballast tray.

We go back to work. Each of us moves within deep walls of thought. Innocence has departed.

In Miami there is more work to be done because of our unexpected failure. A new rubber hull is ordered for SPID, and *Sea Diver* is hauled up on the way for her annual refit. We work in a shipyard on the Miami River, surrounded by large boats in various stages of repair and construction. Under Link's stern eye, we continue to prepare and test the hundreds of pieces of equipment needed for the big dive.

One morning our new deck chamber arrives from New York. It comes to the docks on a long flat-bed truck, covered with dust. Within hours it has been slung in a wire harness by the shipyard's crane and placed on *Sea Diver*'s stern deck.

The squat green chamber is as handsome as a compressed mushroom. However, in many ways it is the most essential of the large pieces of equipment. Within its thick steel walls the two divers will decompress after their deep dive. Although only as large as a Volkswagen car, it will provide essential features for prolonged and safe decompression. Pressure and breathing gases can be critically controlled. Two bunks can be folded down for sleep. Food, water and toilet facilities can be made available. One wall holds a pressure lock where small objects can be exchanged between the chamber and the surface. On the other is a round entrance lock which permits a doctor or diver to compress and join the main chamber's depth. Opposite the bunks is a hatch which leads to the mating collar of the submersible chamber. Two thick plastic viewports allow us to watch the men inside. A communication system permits us to talk freely. It is simple and spartan, but the kind of shelter that ensures the life of its occupants.

A few days later a young man arrives whose life depends on the integrity of the deck chamber. Jon Lindbergh. He has made many dives in most of the world's oceans as a commercial diver and U.S. Navy demolition expert. He is unbelievably shy, but he has a quiet, almost painstaking way of insisting upon answers. Jon's arrival brings fresh insight. We are beginning to feel the fatigue which comes from working too long and too close to a difficult task. Jon's enthusiasm is infectious.

As the big dive approaches more and more equipment continues to arrive at the ship. It is almost impossible to walk across the deck and onto the shore. Extra help arrives, including Jim Dixon from the University of Pennsylvania. The pace moves up several notches.

There are many systems tests to complete before we can hope to dive. One day Link gathers us in the wheel house.

"We've still got a long way to go. Each individual piece of equipment seems to be working well, but its how they work together that concerns me. Today I'd like to make a 400-foot dive with Jon and Robert inside the submersible chamber. I know it's going to be hot in the sun, but it'll be a good on-deck test of ourselves and the equipment."

Link was right. It was a torrid day. The summer sun flamed into the shipyard and slowed every perspiring step. We did not complete our pre-dive preparations until noon. It was obvious that the dive would turn into a fiery sauna. Yet it was important to go ahead. What it was like to be in the metal chamber that day can only be described by a participant. Here is the sensitive pen of Robert Stenuit.

"All morning long under a burning sun I worked in the submersible decompression chamber to connect up the equipment. We were forced to remove the white protective awning which shrouded the cylinder from the sun . . . because we needed access to an incredibly hidden through-hull connection. When I touched the aluminum it burned my fingers. Its interior was a Turkish bath, and we wallowed in the perspiration which poured in rivulets from our bodies. At 11 o'clock, Jon locked 'b' hatch and the doctors increased the pressure. Initially, they bled in one atmosphere of oxygen and then eleven of helium. It was a fast compression, so that our exposure to the gas could be short and the decompression brief. I watched the interior thermometer. I saw the needle start almost immediately from its 86°F. which was the outside temperature. It veered up toward 120°F. I was gasping, and because I was sitting in front of the gas inlet I received blasts of fire right in the face. Soon the needle hit 128 degrees. I had stopped perspiring. I felt my brain was melting. At every breath I could feel new waves of flame penetrating the recess of my lungs. My legs were wobbly. I decided to ask the surface team to slow down compression . . . but what the hell, we were almost at 400. Now we were there. The reverberations of outside voices came to us as in a dream from some other world. With trembling hands, I completed the program; test the pumps and the hookah with the x valves, then with y valves inflate the suits, switch scrubbers on and off. I made notes in a sweat-soaked notebook. Then it was time for decompression.

"Temperature decreased as the gas was bled out. I felt alive again. Thirty seconds, a few degrees drop, and I was in top shape. My pulse, at the worst time, had been 120 per minute, and that, I said to myself, is bad, for my blood will have carried dissolved gas at an increased rate and will have penetrated deeply into the tissues of my body. Before crawling from the aluminum cylinder into the deck chamber, I informed the doctors and they prolonged the last decompression stops on pure oxygen. But time was to prove that this was inadequate.

"There was no mishap as the sun dropped that afternoon. In the evening, between nine and ten o'clock, an acute pain flared up inside my right ankle. It died down, but I went to bed perplexed. A mistake. At four in the morning a fierce pain woke me up which now burned the leg from the knee to ankle, and I had to wake up Joe and Jim, as well as Ed Link and Dan Eden and we went to the deck chamber for treatment. The treatment worked well, but from now on I would be plagued with concern regarding the outcome of the deep dive."

On June 15th we depart Miami. Our destination is the Bahamas and Great Stirrup Cay. As we slip downstream on the full force of the outgoing tide I climb up into Sea Diver's crow's nest. Tiaras of lights crowd the seawalls of the harbour, making diamond contrast with the distant blackness of the sea and sky.

Occasional freighters move their lighted hulls up the Gulf Stream searching for the currents that will hurry them north. I can almost feel the city releasing its grasp. Soon our ship will be free and autonomous on the endless surface of the sea. In our voyage to the Bahamas we will follow the long-practiced harmonies developed between mariners and the sea. But once at our destination we will try to add a new perspective to the relationship between man and the ocean. We will challenge the dark depth dimension. It promises to be a novel and perilous confrontation.

Descent Four
Man-In-Sea
1964

I stand on the ocean floor, seventy feet down. Around me is the soaring blue gloom of infinite wilderness. High on the silvered surface is the black shadow of *Sea Diver*. It looms high overhead like a dark thunder cloud. Beside me is the stout, round shape of the small manned station. It seems dimensionless against the blue obscurity.

My diving partner is Clayton Link. We have descended together to inspect the superstructure and ballast crib of the station. During last night's storm heavy winds shook *Sea Diver* and moved her moorings. A deep furrow in the sand indicates where the station has been dragged by the scudding ship.

Clay and I hang easily in the soundless blue sea. Our initial examination of chains, lines and material has revealed nothing. We are silent, watching the gun-barrel shape of a large barracuda. He looks back at us with the toothy sullenness of a briny building inspector.

We watch him circle slowly, a living silver threat; large teeth jawing the water. His nerves are live wire springs, his course slightly erratic.

We turn as he circles. We know his game. He is shadow sparring with us; amusing himself by swimming around two fleshy forms hung neatly under gas bubble umbrellas. Air escapes easily from our regulators and we are not concerned. We enjoy the game.

The eye-to-eye is a welcome diversion. For the past week both Clay and I have been eager to quit the hot deck above. The work-pace on the ship has been devastating, the sun an indomitable monarch. Any excuse will do. Whenever possible, we drop in to enjoy the cool sway of the sea.

The barracuda turns full face towards us and flicks pea-button

Launch and recovery
of the life-support
systems for the longest
deep dive to date.

eyes. A faint arch appears in his tail. Suddenly he unlocks and springs the coil of hysteria. His silver form cuts the water like a loosed spear. He drives directly between us, jaws agape, teeth able to tear shreds in a throat.

Like frightened squid, we tuck our legs tight to our bodies. A hidden reflex snaps and we compress as small as possible. In a trice we are compact forms hiding behind face masks. The barracuda convulses and disappears.

Clay and I look at each other in amazement. The rules of the game have been broken! It is a match often played between divers and the predator. You keep your distance; I'll keep mine. The big silver shape always stays the same distance, as if tethered on an invisible wire. A feint in his direction pushes him back to a new outpost. Then he glides in to recheck your stance. It is a covert contest and no one ever gets hurt.

Clay and I shake our heads. Something has upset the fish. Strangeness is adrift. Perhaps it is last night's storm.

I duck down to enter the station. Clay stays slightly behind to cover me. I grab the heavy lead collar of the entrance tunnel and pull myself up.

My head breaks the surface. The air in the tight little room feels stale and warm. Water dripping from my face mask has a hidden grotto sound.

I reach for the rubber floor edge and pull myself up into the main living area. The room seems airless. The white curved walls are bathed in condensation. I cannot stay long, for the carbon dioxide scrubber is not hooked up.

It is a small room. It reminds me of a two-man mountaineering tent. Two tiny port holes admit soft blue sea light. The entrance tunnel adds faint gleams reflected from white sand. It is like sitting in moonlight.

Clay's lithe form fills the tunnel. He lifts his mask and laughs.

"That old charger was a mite strange. Wonder what got into him? He's probably hung over from last night's waves."

"How does everything look?"

"Fine. Nothing seems damaged as far as I can see. The battering must have been restricted to the surface. Looks like the station just got dragged around a bit."

"Okay, I'll make one more swim around the outside and check things again. Then we'll head back to the surface. Sure hate to leave the quietness of this place."

"Roger. I'll run over things again in here. Be out in a few minutes."

Clay's tanned legs fin him away from the entrance tunnel. I turn toward the small room and its sparse furnishings. Two white metal and canvas bunks hang against one curving wall. They are folded flush to allow more working space. A small bag of tools lies on the floor. The bag is filled with assorted wrenches, pliers, and two large vise-grips.

I look carefully over the rubber walls and give special attention to the vulcanized seams and portholes. No cracks or other signs of storm damage are visible. I quickly check the other contents of the dwelling including the closed circuit television camera. All seems well.

I see Clay's shadow working its way past the view-ports. I moved down to join him. The water is cool to the skin touch. I shiver and drop down.

The air from my regulator tastes surprisingly sweet, compared to the stale atmosphere of the tent. I remind myself to ventilate it well before the next dive.

The barracuda has returned. He hangs above us like a thin silver satellite. He does not move.

I slip out of the entrance tunnel and steady myself on the station's lifting frame. Clay is waiting, poised like a statue ten feet off the sea floor.

A third diver has joined us. Bates Littlehales, a *National Geographic* photographer, is in position slightly above the fish. He is taking pictures with the barracuda in the foreground and the station in the background. The barracuda cooperates reluctantly. He is concerned about the third manfish.

It is time to return to the surface and the steamy breath of the sun. We three swim upward in silence. Single file, following Clay.

It is an easy, graceful, slow-motion ascent. We have no decompression obligation and can return directly to the hot calm above us.

We swim easily, uneager to shed the cool blue robes. The expanding air from our lungs cascades like an upward niagara. Bubbles are everywhere, spilling, breaking, bursting in abandoned effervescence. The barracuda fades back into the wilderness like an old birch log slipping into a stream.

The sea and the ship's white hull are in a straight line. The water is like oiled glass. A long, rusty strand of sargassum weed hangs limply under the surface. We float easily over to the boarding ladder.

The steel rungs are hot on my hands. The sun has squeezed all coolness into the sea and drapes a heavy warm blanket across our shoulders. We climb the ladder and join the envious crew in the canvas shade of the stern deck.

A new ship has joined us. Her upturned sheer and cabin just aft of centre confirm her as a reconverted shrimper. A grey boom hangs lazily over a bright coloured awning.

She belongs to John Perry, who uses her to carry one of his growing fleet of small submarines. She is *Sea Hunter* and has just sailed in from West Palm Beach. She stands by to assist us with accommodation and general support.

A third ship lies on the near horizon. She is much larger and has harsh grey lines. She is the U.S. Navy's *Nahant*, a net tender from Charleston, South Carolina. Her bow arches into twin gargoyle arms used to recover anti-submarine nets. She seems ominous and somehow out of place on the light blue sea, for she is dressed in battle paint and white-capped sailors move along her decks.

Ed Wardwell is Link's operations manager. As we watch the new ship he describes her function. "The *Nahant* will provide invaluable support. In her holds are stored a hundred and sixty-eight high pressure cylinders of oxygen and helium. This enormous supply of gas is needed to pressurize the submersible chamber, deck chamber, and underwater station." Ed is a former Navy man. He has years of experience with the sea. "At 400 feet we will need thirteen times the volume of gas used at sea level. We have a 50 per cent reserve supply in the event of a gas leak."

The *Nahant* moves slowly away. Ed continues. "She is making a grid survey of the proposed deep dive location. Part of her task is to carefully profile the bottom and select a flat area at the appropriate depth. Once found, her crew will lay out four large anchors at the corners of a square."

The anchors, and their big orange marker buoys, will serve as a four-point moor for *Sea Diver*. The station and the submersible chamber will be connected to our ship by long and bulky cables. Any severe movement by *Sea Diver* will cause the bottom structures to drag and be damaged. The four-point moor will minimize the risk when Jon and Robert are inside.

We are anchored one mile north of Great Stirrup. The large islands look like flat green emeralds. White pockets of sand straggle along the shoreline. At the far end of the islands the pale finger of a lighthouse lifts bleached and lonely in the sky.

To the east and west lies the hidden temper of the reef. It's a gnarled forest of coral fists, washed in snowy lace.

I go back aft to work with Robert who is repairing a slight tear in the shoulder of his new diving suit. It is an experimental model, made for him by the same company that built the rubber tent.

I enjoy watching Robert work. He is always completely absorbed by the task. His hands move slowly and with care, but his mind seems elsewhere.

What is he thinking of? Is he dwelling on the dive to come and its pageantry of risks? Is he thinking of Europe and home? Wherever his mind, it works quietly behind serious brows.

Robert is an unusual person. He has the typical qualities that make a good diver; independence, self-reliance, and the soft glow of rebellion. He is quiet-spoken and pursues a career that suits his non-conformist characteristics. Diving is only one of his talents. For him undersea exploration is a means to explore both himself and the history of man. Few people know it, but Robert's real skills lie in writing and marine archaeology. His consummate energy in both have earned a world-wide reputation.

Unlike most divers, Robert is acutely sensitive to the dangers which face him on the deep dive. He knows the long hazard litany by heart. He chooses not to ignore it—but to surround himself with meaningful activity. Like all men under stress, action helps him relax, especially if it brings the goal closer.

As we work, my own mind keeps drifting back to the litany. I see myself underwater and outside the rubber tent—only this time in Robert's place and under four hundred feet of water.

How would I feel? What would it be like to swim around under that weight of water? Would pressure and cold affect my body? What about sea life and sharks? Will animals be bigger and more aggressive at this depth?

Robert and Jon will spend much time inside their small rubber room. I wonder how cold they will get, and their reaction if their breathing mixture fails. Without oxygen the brain shuts down in less than five minutes.

Robert finishes his repairs and beckons me down into the diving locker where most of the team has surrounded the big air compressor. On this hot day the locker is dark and almost airless. Its shelves hold a line of silent scuba bottles and swim fins. A gang of muttering men has gathered around the compressor. They look down on it with the temper usually reserved for street fights.

They are not happy. The collective mood is only a linchpin away from giving the uncooperative compressor swift burial at sea.

"Damned thing. Turns over once or twice then quits like a belch going upwind."

"Why don't we push it over and use it as an extra anchor. Or maybe ballast for SPID."

"Yeah. Son of a bitch has as much compression capacity as an old hernia."

The air wobbles with phrases found in a Shanghai tavern. Then silence. A dark head swings in from a partially open hatchway. It is Jon Lindbergh.

"Let me have a look at it. We've got one of these cantankerous old billys at home. Maybe I can. . . ."

Sweat-soaked forms step back and then bend over to watch Jon work. He goes straight for the electrical system and begins to open and close a series of switches. He stands up.

"There. Let's see if we can get some breath out of this donkey."

Someone pushes the wall mounted starter button. Nothing. Again. Electricity flows. Resistance. Nothing.

A large form, naked from the waist up, steps up in front of the inert machine. He glowers at the compressor.

"I once worked with a tractor that had your same bloody manner. A slight touch of the hammer usually brought her to heel."

He bends his thick piano leg and kicks out like a mule. His foot lands squarely above the crankcase.

The motor responds. A slow whirring, and the compressor chuffs to life. Acrid blue smoke swirls in through the open hatch above us.

Robert turns to me. "You can't be too kind to these things. Otherwise they'll take advantage of you."

The team drifts back to their various jobs around the ship. Most of my attention is given to the general life-support systems for the deep dive. Safety and efficiency are the key words. At times I treat lacerations, bruises, and other injuries. All of us know, however, that our onboard activity is not limited to any particular field. In moments when my own work is not demanding, I fill in wherever an extra hand is needed. And in times of emergency all available men are called.

One night, for example, I was wakened by a soft knock on my door.

"Better scramble, Doc. We need an extra hand topside."

Lightning flashed through a porthole. I reached for my jeans and then dropped them. Rain gear and a bathing suit are a better combination. A caucus of fire hoses seem to be splashing the deck overhead. I slipped on a pair of boat shoes. The steel would be slippery.

I sprinted through the dark ship and up the companionway. Sightless darkness covered the ocean. Rain in primal gloom. For some reason the ship's deck lights were not on. Perhaps the storm.

With one hand on the storm rail, I made my way to the stern. An avalanche of thunder tore the sky. Two figures were huddled over a huge coil of cable and hoses—part of the umbilical that connects *Sea Diver* to the station seventy feet below.

The umbilical was caught on something on the deck plates. The storm was already putting a strain on the long length of cable already in the water. I pictured the station being towed across the sea floor. A slalom over the sand. If she hits something solid, she might break up and destroy herself.

The faces of the two men shone like wax in the dark. Tears of rain raged down their faces. They work over the low fire of curses.

"If we could just get some damn slack, we could free it."

A fourth man joined us. It was Ed Link. Together we heaved and pulled on the huge coil. No use. It stayed hung on the hidden grasp.

Suddenly the sea reversed itself. Waves started to smash the other side of the ship. The wind and sea pushed us slightly towards the umbilical. It began to go slack.

We quickly found the cleat where the umbilical has seized the deck. The umbilical's free end, attached to a buoyant white recovery float, was tossed over and lost in a wave.

As we made our way toward the galley, Link spoke out over the rain.

"A good job, men. I guess our starboard anchor dragged. Good thing we had the umbilical buoyed and ready to toss."

A faint smile crossed his face. He was warm and breathless from the exertion. Another crisis met and defeated. We are all bonded even closer by these confrontations.

The days slip by. Long, hot periods of calm are interrupted by squalls of angry rain. Sudden black storms rise out of the Gulf Stream and beat angry wet fists against our small flotilla. I am concerned. The deep dive needs a sustained breath of good weather.

Our seventy-foot test dive with all systems is successful. An essential element of confidence has been added. We step up to the final threshold.

Sea Diver slips into her four-point moor. Over four hundred feet of water lies under her keel. The small rubber station is inflated and its ballast filled with lead. It disappears into the blue shimmering sea followed by a trio of pilot fish. We track its descent on closed circuit television.

On the morning of June 30th everything is finally ready. The sea is calm and the weather perfect. *Sea Diver*'s decks are extraordinarily quiet. We move as grooms just before the wedding.

Everyone's attention is silently focussed on Jon and Robert. In a few minutes they will begin ascent of their underwater Everest. They will surround themselves with the cold snows of danger.

The best view of such an experience can only come from the inside. Let us turn again to the sensitive pen of Robert Stenuit. The words are as he wrote them for the *National Geographic Magazine*. The italics will relate some of the impressions we had on the surface.

June 30th, 0945

The sea is calm, almost as blue as the sky. Ed shakes hands with us and wishes us good luck. It is my last sight of the surface.

We go under and swim to the bottom of the cylinder, which stands vertically in the water alongside *Sea Diver*. Jon opens hatch "a," and swings it outward. We climb in. I close hatch "a" to prevent water from getting in, then the second hatch, "b," which opens inward and prevents inside gas pressure from getting out. With both hatches shut we have locked our doors against the outside world. For six days we will not again breathe God's fresh air.

1045

At the control panel aboard *Sea Diver* Jim Dixon slowly increases our interior pressure to equal that of 150 feet of sea water. We will hold that pressure all the way to the bottom. Now we are breathing 77 per cent helium, and it shows its effects . . . when we speak, we quack. Jon handles the telephone; a Donald Duck voice will be more easily understood topside without my French accent.

1150

The depth indicator shows 60 feet. Through the port I see a diver working on one of our electric cables. It is Ed Link. From first to last he has insisted on checking everything himself.

1215

Our descent is so slow that only the creeping depth-indicator needle reveals it. We have been ballasted slightly heavier than water so that we will sink. Braked by a safety line from the surface, we slide down a nylon guide rope that angles gently to the SPID. When the anchor hanging below us touches the bottom, we will come to rest, floating upright six feet off the sand.

1230

Three hundred feet. The water is limpid, but it grows dark. Joe MacInnis phones: "We'll be taking samples for a few minutes to check your gas mixture."

> *The incredible distance of pressure and depth begins to separate the divers from the surface. Only voice and television connections close the gap. I can now sense the emotions of an astronaut doctor as the spacecraft takes off.*

Even though our cylinder is autonomous (we have our own gas analyzer, carbon dioxide remover, and oxygen supply), a backup system can feed us gas and control pressure through hoses from above. The doctors test our breathing mixture periodically.

1234

"The SPID!" Jon has sighted it at last, well-placed on an even bottom. It warms the heart, this view of our little house set in a landscape that is more lunar than terrestrial.

1300

"On the bottom." A big smile lights Jon's face as he reports our arrival. The depth gauge at the top of the cylinder reads 415 feet. Add 11 feet, the height of the cylinder, and six for the chain between the cylinder and the anchor: 432 feet. . . .

> *I am envious. We are envious. We are chained to the surface—you are about to fly free in the ocean.*

1315

Through the long umbilical cord that connects us to *Sea Diver*, the surface crew sends us helium to increase the pressure in the cylinder to 14 atmospheres—which is the pressure of the water outside. Then we will be able to open our double doors easily. In perfect balance, gas and water will meet at the entrance to the cylinder.

Our breathing mixture is now a cocktail of 3.6 per cent oxygen, 6.6 per cent nitrogen, and 90.8 per cent helium. The helium, under these pressures, prevents us from producing intelligible conversation. I scribble messages to Jon on the wall.

1330

Our diving suits are flattened and wrinkled like old parchment. I blow them up with a flask of compressed air.

1345

Word comes from the surface: "Okay to open hatches." I pull hatch "b" upward, push hatch "a" downward. There is the Atlantic, a circular patch of clear blue lapping at my feet. I let myself slide into it, and shiver. It feels cool, even at 72 degrees Fahrenheit. Visibility is more than one hundred feet, despite the twilight gloom at this depth.

I glance around, looking for the big sharks that we have been told to expect. Nothing in sight. Our outside spotlights pierce the gray water with two emerald beams and awaken glimmering splashes of sleeping colour on the sand. Our shelter looks all right. Its hoses and cables soar upward toward the hidden surface. But it is a bubble in the immensity of this foreign place.

It is only 15 feet away, close enough for us to swim to it without breathing gear. I glance upward. At this depth we could not hope to struggle to the surface alive. But we have a return-trip ticket: the cylinder waits to take us home.

Seen from below, the water surface inside the entry shaft of SPID is a mirror of blue silver. My head breaks through it. The gas of the interior tastes like fresh mountain air. I climb the ladder. At last, at last, I am here. Six months of delays, of dogged effort, but now I am here. What calm there is in this other world. What silence. What peace. I shake myself. I must act quickly:

My first task is to connect the gas analyzer to its waterproof batteries. The little black needles come alive: oxygen, 4 per cent; carbon dioxide, 0.025 per cent. All is well.

To avoid damage in case the SPID should be flooded during its descent, all the instruments, the electrical connections, and the interior equipment have been enclosed in waterproof containers attached to the ballast tray beneath the chamber. We must get them unpacked and installed.

1405

Jon has joined me. First he hooks up the wires that will establish contact with the surface. Afterwards he connects the light. The light bulb glows, burns five seconds, and goes out. We look at each other in consternation. Is it the light bulb or the current?

We continue to work in the light of a diver's hand lamp. A noise like a gunshot slams against our eardrums. The sealed-beam bulb has imploded, spraying the inside of SPID with thousands of sharp fragments. Happily, we still have flashlights.

> *We detect the trip-hammer ramming of your hearts. Your coolness is amazing; your voices do not betray you.*

I plug in the radiator. Nothing, without heat for our atmosphere and our food, things present themselves somewhat poorly.

1410

Standing up in the narrow access well with water up to my waist, I wrestle with a four-foot aluminum cylinder. It houses a machine that will filter the gas in the SPID and remove the excess carbon dioxide. At last I get its top out of water and open the equalization valve.

This valve allows pressure inside the container sealed at sea level, to equalize with the 14-times-higher pressure inside the SPID. When the valve is opened, gas should rush into the container with a loud psssst. Only then can the lid be removed.

I turn the valve, but nothing happens. The container is full of water. Catastrophe! Our situation is definitely not brilliant; the apparatus is vital to us.

On the surface we can tell that things are going badly.
The silence is appalling.

I glance at our analyzer and see that the carbon dioxide level has risen to 0.1 per cent. Our minutes here are numbered. Quickly we fetch the spare filter. It seems to weigh a ton as I thrash around on the bottom, trying to drag it behind me.

Jon hands me a line. I push and he pulls; I lift and I pivot and I manoeuvre. I come back to the entrance well to breathe more and more often, more and more heavily. At last the monster is in place, but I am completely out of breath. Our furious efforts have raised the level of carbon dioxide to 0.17 per cent.

Now we discover that this container has no pressure-equalizing valve. The surface people have put on the wrong cover. I calculate rapidly: about four tons of pressure hold that cover on. No use trying to force it off.

A shock wave runs through me. I put that cover on. In the last minute rush I made an incredible error. My spirit collapses like a bag of loose sand.

Can we pry it open enough to let air in under the edge? No luck. I break a screwdriver, and Jon snaps a scissors blade. A glance at the analyzer shows 0.2 per cent carbon dioxide. We are panting now, breathing too fast. The heavy pounding of my heart resounds through my whole body. I make a sign to Jon: Get out. And we return to the cylinder, to the sure refuge.

1430

I consider our condition. Without a gas purifier, without light, without heat, perhaps without any electricity at all, it is discouraging. I write with a grease pencil on the side of the cylinder: "In any case we will stay here 24 hours."

From the level above me, Jon signals his agreement. An entire day spent below 400 feet in the cylinder would be at least a halfway success.

We report our predicament to the surface in morse code. Link answers efficiently, as always: "We are sending you a line. Attach it to the flooded container. We have an exchange motor. We will repair and return."

1600

We wait. Dr. Dixon calls us. "According to our instruments, there is now more than 0.2 per cent carbon dioxide in the SPID. You have a maximum of 15 minutes inside."

When we re-enter the place, the carbon dioxide level will climb very quickly. About 0.5 per cent, its toxic effects will be severe, and there is no rescue possible. Those 15 minutes will decide the success of the entire operation.

You can't know how hard we are working up here. Recover the flooded motor. Repair it. Return it to the sea. I am still numbed by the scale of my blunder.

1700

We wait. It is growing dark.

1825

Something clangs against the cylinder. The new gas purifier has arrived. I leap into the water, shivering, and drag it over to the SPID. Inside, the gas seems heavy and thick, sticky in the mouth.

I open the equalizing valve. This time gas rushes into the container. The purifier is dry. Jon's face lights with joy, but we have no time to celebrate. Six minutes have passed already. The analyzer's needle creeps into the danger zone. We take off the cover, wrestle the machine into its cradle, and plug it in. The motor purrs. The gas circulates. We have won.

1930

We are installed in SPID. Tonight's dinner: carrot juice and corned beef, canned water, fruit salad.

The relief of success washes over Sea Diver. *Our confidence returns. We settle into the rhythm of the watch. I make a deep mental note of the lessons learned.*

2300

I have taken the first watch of our first night at 432 feet. I keep my eye on the instruments and the level of water in the entryway. The radiator does not work, and Jon shivers on his cot in three sweaters.

On the surface the medical team keeps vigil in relays, scanning their gauges and observing us through the closed-circuit television. "Big Brother" is watching.

You both look so damn cold. Two figures huddled in a fireless tent. Yet you say nothing. The courage is impressive.

0200

I lean over the wall, and my heart suddenly rises. A huge black silhouette moves slowly against the ladder. A shark? No, it is a peaceful grouper, as big as a boar.

0900

Breakfast. As soon as we move, the temperature becomes bearable.

1000

To work. We drop down through the well and into the open sea. We test our breathing apparatus. We have no back tanks, for they would only last moments at this depth. Instead, we use a "hookah," a 50-foot double tube which feeds us breathing mixture from the SDC and carries off our exhaled breath for purification.

Jon swims around at the end of his hose, exploring the coarse sand bottom. We see life everywhere: sponges, worms, anemones, octopuses, and minute, royal-blue fluorescent fish which I would like to catch and make into rings or earrings.

The big grouper follows us everywhere, nibbling at my feet when I come down out of the access well. He accepts all our caresses.

1800

Jon has repaired the radiator and dehumidifier. After hours in the water, it is pleasant to return to a warm and dry haven.

2200

We try an experiment with voice communication. Question: Below what percentage of helium can we make ourselves understood at this depth? Jon breathes three deep gulps from a bottle containing 25 per cent helium and 75 per cent air. His voice remains nasal and deformed, but I understand him clearly and the surface does too.

He dictates telegrams to his four children: "We are in a small rubber house on the bottom of the ocean. Hundreds of little fish are swimming outside the window. . . . Two little octopuses were playing on the bottom under us yesterday. They would glide into a hole and then jump out at the fish. The fish darted away, but always came back to watch the octopuses. Then we swam out, and they all ran away. . . ."

The radio on board *Sea Diver* relays to the four children a fairy story become a reality.

Now it is my turn. I try three deep breaths of pure compressed air. The air is so dense that I can see it flow out of the regulator like a thick fog. My voice takes on human tones, but at the third gulp the SPID begins to undulate. I feel my face twisting into ludicrous grimaces. I am drunk. I let go of the mouth-piece. I can do without nitrogen narcosis.

So can we. You scared us for a moment. Fortunately your recovery was equally rapid. We all breathe easier.

2315
Our last 1,000 watt exterior spotlight, which burned night and day, goes out. Our interior light brightens immediately. Ten seconds later the water in the access well is boiling. All the little fish attracted by the big light have now come around to the lesser one. They twist and turn and jump out of the water like mad creatures. At once I see why. The water is alive with tiny shrimp.

2335
A heavy blow shakes the SPID. Jon wakes up, startled. What is happening? Another shake. We hold on to the cots. It is the giant grouper, charging the fish in the well with his enormous mouth wide open. Ten times during the night he awakens us.

July 2nd, 1000
We go out to spend three hours in the water. I take pictures of Jon working on the SPID. The light of my flash attracts half a dozen giant groupers, but I cannot photograph them. They press against me, nudge the lens, bump into my legs, fill the entire field with their bovine bodies, and stir up sand with fins like ping-pong paddles.

I tore my diving suit yesterday from shoulder to waist, now my teeth chatter helplessly. But I must continue, frozen or not. If I succeed, my pictures will be the deepest pictures ever taken by a diver. When I cannot stand it any more, I return to the SPID and greedily swallow six big spoonfuls of sugar. Thirty seconds later I stop shivering.

1320
The surface is calling us. "You have spent two days and two nights at 432 feet, and all our tests have succeeded. Bravo! We will gain nothing more by extending your stay. Prepare to come up."

We look at each other. Now that we are installed and well organized and have made friends, we would willingly stay a week in our little house. It is the voice of reason, however, and we obey.

A difficult decision. Especially for Link. But the weather looks ominous and it will serve no larger purpose if you stay much longer. Secretly we all want to see you safely on the deck.

After that comes the routine of going up again: the elevator sealed, the needle of the depth indicator coming down across familiar figures, the water growing lighter. Then the cylinder dances on the surface, and we are hoisted aboard.

Happy faces peer in at us through the port-holes. The cook announces a steak dinner to celebrate our return.

Our cylinder is joined to the larger deck decompression chamber with its two cots and its air locks that permit supplies to be passed in to us—and a doctor, when necessary. Then we drink our first iced drinks, eat our steaks, and crawl into our cots. I stay in mine so long, and return so often that someone suggests changing the name of our project from "Man-In-Sea" to "Man-In-Bed."

Months of weariness have brought victory. It's the first good sleep you've had in weeks.

Project Man-In-Sea goes well. Ed Link has triumphed again. We have stayed, Jon and I, longer at greater depth than anyone before us. We have set a new record. But what is more senseless under the sea than a "record"?

What we have accomplished is something else. We have lived in the depths—in questionable comfort, but at least in security. We have gone out and worked. To be sure, we paid for our two deep days with four days of decompression. But if we had stayed two weeks or two months, the decompression time would have been the same.

Our successors will stay in the depths that long and longer. They will colonize the sea floor, cultivating its resources instead of pillaging them. Tomorrow the colonist will survey his bottom land through the porthole of his sea-ranch kitchen while a coffeepot simmers on the stove.

Descent Five
One Hundred Fathoms
1965

I sit waiting. The edge of fear joins me like a stealthy moonbeam. The taut canvas bunk below me is cold and its steel pipe frame harsh against my skin.

My chest itches. Two round flat sensors have been glued over my heart to detect its rate and wave pattern. I can sense its quiet gallop under my rib cage. The curved white walls of the compression chamber glisten with sweat.

In a few minutes the thick round hatch beside me will swing shut and the dive will begin. I try to suppress the anxiety that keeps putting its cold hand into my lungs. My breath feels arctic and thin.

This is no ordinary dive. My companion and I are going to compress to a depth where only a handful of men have gone before us. 650 feet. The palms of my hands dampen with moisture.

I am surrounded by the tight steel womb of technology. The ribless walls of the chamber are one and a half inches thick and cold to the touch. Their iron integrity is only interrupted by two small round viewports. Through the thick plastic lenses I see the flicker of distant fluorescence.

A communications set in front of me carries the remote voices of the surface team. They stand on the other side of the rigid steel, but seem a thousand miles away. Beside me, a wall-mounted control panel clusters brightly with gauges, valves and dials. Below it, near the floor, are the ominous outlines of two moss green masks. They belong to our emergency breathing system.

I put my hand up and run it lightly over the end of a one inch pipe. Its onyx opening will soon admit the gas that will compress our bodies.

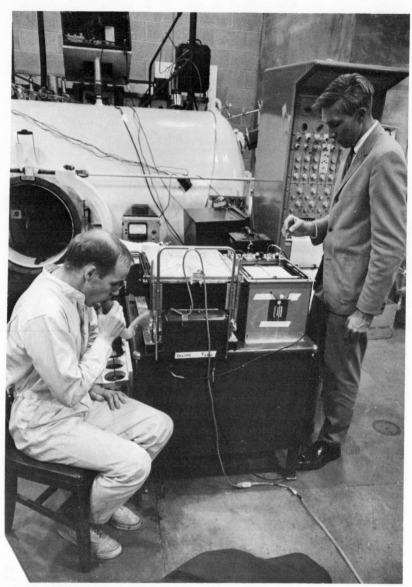

The author (left) just
prior to 650-foot dive,
blowing through a respirometer.
The pressure chamber
is in the background.

This is a laboratory dive. It will simulate all the conditions found in the sea except cold and wetness. It is an essential step, for it will tell us much about human reactions on the bottom, and the success of our newly developed decompression schedules. I have volunteered for this dive of the series because I am anxious for the first-hand experience. It is the only way to be certain what actually happens. It is the difference between knowledge and speculation.

As the seconds tick by I watch and wait. The phantom of last-minute reluctance appears and joins me in the chamber. I wonder: What am I doing here?

My companion is Bev Morgan, a commercial diver from California. Bev is an unusual explorer: he dives, invents equipment, and writes about his experiences. He sits beside me looking ahead. He does not smile.

My thoughts move sluggishly down a narrow endless tunnel. On the surface I look calm, but rumbling through my head is a quiet congress of concern. My mind flicks over the dangers like a shark seeking prey. Is the chamber pressure tight? How secure are the through-hull fittings, and their root-work of pipes? What if a viewport should fail at 650 feet? I picture everything in the chamber, including Bev and I exploding in a pressure stream through an eight-inch opening.

I carefully attempt to put away each irrational thought as soon as it arrives. I feel like a man whose work is continually interrupted by a parade of uninvited visitors. I slam the door, but the thoughts continually push it open.

The impotence of waiting begins to press in on me. I do not know it, but the relief of well-rehearsed action lies just around the corner.

The hatch swings shut. Its ivory coat of paint is unsettling; it looks so clinical and pale. I am not excited by the fact that it is also fireproof.

Silence. It washes Bev and me in long, ill-defined waves. We know the team is moving around outside our steel cocoon, but we cannot hear them. My heart moves into faster calibration.

"Let's get this thing moving," says Bev. He too seeks the comfort of meaningful action.

"If you're ready, we'll begin the countdown." The electronic message has authority. The topside team is poised. Three hours of final preparations and endless checklists are over.

"Roger, we're ready. Let's go." I try to say the words with conviction, but they arrive without muscle. My mouth has a slightly metallic taste, as if I'd just put my tongue on warm brass. The electrocardiogram detects another rise in my heartbeat.

"Five, four, three . . . two . . . one." The air explodes with the waterfall rush and roar of helium, blowing in to pummel my ears like a sound out of control and getting louder. It is a sound like brushfire in the tree tops and makes my mind narrow-tight to the point of life. Helium molecules gather speed, compressing each other in billowing waves of heat on the skin, in the lungs and across the eyes. The tympanic membranes of my ears stretch now, and hurt with the pulse of pressure and noise. The old skill grasps hold and blows gas into the middle ear to find transient relief until it must blow again to equalize the thunder and helium that threaten to tear my eardrums. I am in an autoclave out of control.

"One hundred, two hundred. . . ."

Feet of pressure read to me by some unknown voice struggling through the noise. I try to right my brain from its new position in soft quicksand, and realize that this is not the time for poetry or mathematics or attempts on any kind of intellectual sophistication for the noise—heat—pressure tunnel is too overwhelming.

"Three, four hundred. . . ."

Numbers moving by slowly in time-frozen series. The only activity here and now is a skinny black needle climbing up a snow-white gauge, and the damn helium, a hell-fire furnace, blowing its solar breath right into my lungs so that my alveoli recoil and shrink. My heart pumps faster, in response to the stresses of the new gas environment pushing in at great speed. One hundred feet of compression a minute. Almost fifty pounds per square inch added each sixty seconds onto human tissues not designed to operate outside the narrow envelope of cool air which surrounds the soft green planet and its sea. An atmosphere so different; this sauna and its blistering heat scream right inside the tight ceramic of my brain, making my hands shake from an invisible cold.

"Six hundred."

The noise less now, but sizzling heat still pouring out of the gun-barrel gas pipe pointed at my head. I feel light-headed and dizzy, and my eyes do not track well although they see the vague lines where air, oxygen and helium meet and waver. It's like the shimmer-heat lifting above a radiator in front of a window, or where fresh and salt water mix in boundaries unknown and unmeasured as the searing, feverish wind and its hideous sister noise now suddenly come to a complete and final stop.

"On the bottom. Six hundred and fifty feet." The voice is strong and somehow comforting. Yet it is a long way away, and somehow blurred like night fog on a lake.

The first few minutes of the forty we will spend here have been set aside for "adjusting" and making notes on the six-and-one-half-minute trip that took us to a depth equivalent of the deep continental shelf. My mind welcomes the quiet and cooling like a thirsty animal seeking water. Thoughts begin to run at old and comfortable rhythms. But, my hand still trembles like an old man with an advanced neurological disease.

I look over at Bev. He too watches the uncontrolled quiver of his fingers trying to push away from each other. We both smile and hold out all four hands for shared inspection. We giggle. There is the pale suggestion of long white shorebirds attempting flight.

Our work begins. Bev, looking at me through the steamy window of his own perspiration, gives the okay sign. Thumb to forefinger. I nod.

Bev opens his mouth and begins to speak in a voice suddenly not his own. The words are high-pitched, and utter from the swollen throat of an old shrew. It is a science-fiction voice. Helium gas playing quiet hell with human resonance. We have both heard the discordant duck quacks many times in other people. Today it is us, and we laugh.

Bev reaches down for some exotic ironmongery. Brushed aluminum plates connected to a box with hoses and wires coiling out of it. A respirometer. A device for measuring the rate, force, and depth of our breathing. Bev begins to blow into one of the hoses. He follows closely the instructions given from the surface.

We are now at a pressure twenty times that found outside the steel walls. Three hundred pounds per square inch.

But I am not aware of the pressure. It is so evenly distributed over and through my body's cells that I have no sensation of weight. If I were in the sea I would be under a column of water as high as a 50-storey building. Yet, I am not directly aware of the tower of pressure above me.

Bev works away blowing breath after breath into the machine. A twinge of discomfort crackles through my right wrist. I felt it before we reached the bottom but in the din of the descent it seemed unimportant.

It is a phenomenon that we call "hyperbaric arthralgia". It is a pain related to compression and high pressure, and usually found in the joints. I move my wrist backwards and forwards and feel a light knife scrape of discomfort. A diver once told me, "it is like having no joint juice." He's right. The pain is slightly uncomfortable but not disabling.

I take over from Bev. He monitors my breathing performance and makes notes in his log.

We both are more certain in our movements. The chamber has cooled to a reasonable temperature, and the sound of the carbon dioxide scrubber is soothing.

From some inner recess I feel a slight glow of euphoria. It is not narcosis, for helium exerts no narcotic effect even at these depths.

No, it is something different. It is glandular. It rises from a secret tide deep within and bathes me in the soft wash of well-being. It is the euphoria of achievement suddenly attained; of temporary control of the tightrope. It is a seductive luxury. I put it away. It is for another time.

I puff and blow like a winded moose. My breath disappears through a mouthpiece and down one of the python hoses. A hidden sensor in the mouth-piece detects carbon dioxide levels at the end of each respiration.

There is a bellows arrangement in the box between Bev and me. With every exhalation it sings a chronic metallic chorus. Music from another planet.

Every minute, instructions come down from topside with the slap of static.

"Breathe deeper."

"Keep your eye on the flow loop."

I remove the respirometer mouthpiece. I am dizzy from the effort.

The respirometer feeds its data along slender wires out of the chamber through special fittings. Our physiologist, Dr. Bill Hamilton, sits at an outside console looking at five automatic pens. They make long wiggle patterns on an endless strip of paper. Bev's and my readings are brothers, printed out side by side.

Bill is specifically watching my heart rate and respiratory rate. He is alert for any sudden change. His concern is comforting.

We begin the second phase of our bottom tasks. We exercise by lifting heavy weights from one part of the chamber to another. This work will simulate some of the heavy muscular movements of commercial divers working on a well-head. As we work, we talk. My voice sounds like the shrill of some knife-wielding fishwoman.

"Bev, you okay?"

"Yup. Except my eyes. At first they didn't focus quite right."

"Me too. I also got this niggle in my right wrist. It's better now."

Our lips work vainly around the strange new sounds. Vocal chords are resistant. We speak slowly in simple sentences, but the phrases have the high keen of madness.

We continue moving the lead weights. Pick up, swing, and lower. Pick up, swing and lower. The rhythm is spontaneous.

The sweat begins again. Freshets trickle off our foreheads and drip to the floor.

I wonder about temperature control in the chamber. One minute I am hot, and the next I am cold. My temperature regulating system seems to be confused. My thoughts float over the effects that pressure may be exerting on my brain. What is "normal" for this depth? For sea-level?

The muscles in my shoulders become tired. An instant switch from effortless to effort. Manic to depressive. On-off.

Time to go. Finally. Forty minutes—suddenly unmeasurable.

We don our thick sweat suits and wait for the new countdown.

"Three, two . . . one."

Glances are exchanged between two friends hard in the grip of life-saving friendship as a new sound fills the chamber like a waterfall heard through the leafy forest. Not a huge sound like the first one but a kind of distant issuance. The helium now runs out and up a cold pipe and into the sky above us, finally free in mad molecular escape, while down below two men huddle and hunker down in a snowless cold, seeking comfort. I concentrate on breathing out, reading a sign in magic-marker ink scribbled on the hatch which says, "SING ON THE WAY TO THE SURFACE." The sign is deadly serious. To hold one's breath is to entrap air in the lungs which can only be released in a shower of deadly bubbles into the bloodstream. They can lodge in the brain, block bloodflow, and cause paralysis or death. So I breathe out easily, hoping each alveoli will safely unload its burden of expanding gas, and hoping that my lungs will hold fast and no hidden defects in tissue structure will appear.

A steam of condensation clouds the chamber as the escaping gas rushes up the frosty pipe. Thin vapours tickle my throat and reach down to ignite my cough reflex. It is an urge to be resisted, for to cough is to hold the breath. The hands swing up, cup the mouth and slow the tickle of vapour's insistent itch.

Nothing to do now but sit tight to the bunk, stay warm, relax if possible, breathe easily and let the mind wander in on itself.

My eyes move over to the pipe opening which empties the chamber. The thought of an accident and my skin trapped by the evil suction flickers across my brain. Flesh and blood vacuumed out with surgical deftness. I put the evil away and instead run over the check-list of safety procedures. It is an agenda which flashes like bold neon through each second of the dive.

The shivering becomes violent now as we hurl upward from the deep tomb depth. I hunker down even further into my cradle of cold.

Topside speaks. Prophet tones.

"Your first stop. You've arrived at your first stop. How you feeling?"

"Damn cold, thank you," says Bev. His voice is a little more familiar. The helium squeak has softened.

75

Our first step toward the surface has been a big one. We are at 370 feet. It's taken a little less than four minutes to travel.

From this point on, it will get more tedious. We have about twenty hours of decompression to look forward to. Twenty hours to come back from 650 feet. A time-penalty for forty stolen minutes.

For a while Bev and I sit and talk. We have the animated conversation of two old cronies trying to describe a new experience to each other. Our voices take on the bar-spirit tone found at the beginning of an evening of promise. We curse the heat and marvel amazement at the track of events so far. Then we lie in our bunks and settle down for the long ascent.

Both of us know that the toughest part of the dive is yet to come. A half-formed thought not allowed to mature. Decompression. A complex staircase of time, depth and breathing as requirements that will finally land us on the surface. Initially, the steps are brief, but as we approach sea-level pressure, they lengthen out in exponential fashion. The most difficult part of the dive profile.

It is in the shallow regions that decompression sickness is most likely. We are both aware of its fierce displays. I have seen destructive pain-shells fire through proud young bodies. I recall an old friend who had succumbed to the dark winds of vertigo. A ruthless bubble lodged near his brain. He was in such distress that he threw up. I remembered the hard grey stillness locked in my gut as we nursed him slowly back from the cliff edge of shock.

The two novitiates wait calmly. We trust men and machines. Heinz Schreiner and Pat Kelly have developed a novel method of decompression and have programmed it into an IBM 360 computer. We believe it is the most advanced approach to the problem anywhere in the world. Our topside crew is the best to be found and will follow instructions to the letter.

A bead of worry trickles down my forehead. These are tests. They are the deepest dives anywhere in the world.

Hours pass.

From my bunk I hear the muffled sounds of my teammates working quietly outside the chamber. We have worked together for over a year. Confidence and friendship are tied together in a secure knot. Time-tested. Bev and I are totally reliant on our friends. They know this and respond quickly to every request. Each man knows what it was to be inside the chamber.

More hours pass.

I first notice the pain at 50 feet. It comes like a phantom, touches the inside of my right knee, and then disappears. I move my leg. Everything feels fine.

A few minutes later I feel it again. A light gnawing sensation deep inside the bone. Nothing serious, just the breath of discomfort blowing.

I rub my knee and move my leg. The discomfort disappears.

I try to occupy myself with other thoughts. I have seen men sit in decompression chambers with nothing to do. Their minds roam. They begin to imagine illness. I start to read.

A tiny fire flickers somewhere in the marrow. A spit of flame deep inside the bone. Just like the imagined pain that blooms in the brain before the dentist inserts his drill. Or is it?

The thought hits me like a thrown stone.

Real? Imagined? If it's real, then you better not wait too long, for that bubble and its friends have blocked the blood supply and oxygen is no longer flushing into the cells. The bone may die, to leave you with an island of necrotic tissue. If the bone dies near the articulating surface of a joint, it may break down, leaving you with a painful disability and permanent limp.

I feel the bone sting of an asp.

The book drops to my lap. Reading doesn't help. I am an insomniac, caught in the thrall of cerebral argument.

If it's imagined pain, then you'll bring to a halt this test, these men, and all this complicated equipment for no apparent reason except unyielding fancy. The decompression schedule everyone's trying to validate will be that much more imperfect, and there is the possibility that invented pain will not respond to treatment. So there you'll sit, looking at the "pain" getting worse and no one able to help, as your loosed illusions take darkened wing.

I detect the sullen presence of ego in the background, whispering, "Whatever you do, don't look foolish. Mustn't ever look foolish."

We are near the forty foot stop. I arrive at the cliff edge of decision.

If the pain increases during the next scheduled pressure drop I'll report it to the surface team. If not, I'll read this damn book.

A silent wind issues from the exhaust pipe. Bev changes position in the bunk above me. I look around the chamber savouring its history. It's the same one we used during the Stenuit-Lindbergh dive. Jon Lindbergh slept here. We call it affectionately, "the white whale." The place has all the glamour of an antiseptic sewer pipe.

The pain hits hard like a tooth exploding. Both knees. A pool of acid below the kneecaps. The nerves in my legs are incandescent.

"Topside, I think we've got a problem. Looks like I've been hit in both knees."

"You sure, Doc? We're almost home."

I am sure. The acid spreads with the malignancy of hot lava. I swing my legs off the bunk and try to stand up.

My legs buckle. The pain has boiled them into soft jelly.

"Damn!"

"I think we better turn around and go back down. Hate to tell you but my knees are really hurting."

Shadowed voices crowd the intercom. Cold conversations. Consensus quickly reached.

"Okay, Doc. Stand by. Increasing the pressure at ten feet per minute. We'll pause each ten feet to see how you're doing."

My ears feel the gentle slap of pressure. In three minutes we travel back down 30 feet. A distance that took us hours to cover on the way up.

I feel depressed. I have just added infinity to the schedule.

We stop at eighty feet. Bev sits looking at me. He is concerned.

"You okay, Joe? You look a little pale."

I feel as cold as an old walrus tusk. The only sound is the running drum of my heart. The pain is not going away.

Usually if pressure is increased the relief is almost instantaneous. Usually. Not now. My optimism is gutted. My knees are unchanged. Blood, flesh, and bones cry quietly with pain.

The insomniac argues another dilemma.

Go deeper? Risk more lack of improvement, and a huge time penalty? Or wait?

The sharks continue to hold both my knees with sawtooth jaws. I decide to wait.

The pain eases. Like sun lifting fog from a pond the pain steals away. My leg muscles quiver with relief.

I sigh and lie down. The pain ebbs.

More sounds from above. New people have arrived. Voices gather like gulls following a ship.

"How you feeling, Doc?"

"Better, much better, thank you."

"Good. Now you take it easy, Doc. We can't afford to get you too banged up. Otherwise, who's going to look after us? And believe me, Doc, if you saw the shag-eared group that's outside this chamber today, you'd agree. They sure need some lookin' after."

The comment rides in on a Polish accent. My old friend Frank. He is today's chamber operator. His hands are as quick as his humour. Diving with Frank means you never know when a verbal bite will echo over the intercom.

Frank continues to lay his laughter over the dark situation. Standing beside him is a small group of men quietly outlining the fate of my decompression. They include Hamilton, Schreiner, and Kelly. I have been part of this same group for other "hit" dives. There is a set routine.

Depth and time of the accident. Gas mixtures. State of the diver. Degree of his response. New depth and time of relief. These and many other factors are reduced into numbers and estimates. The computer and the original matrix are consulted. A new decompression schedule with longer stops, slower ascents, and different breathing gases is constructed. The return journey is resumed.

Suddenly I really understand what it is like to be on the "other side of the steel." I have always been part of the topside team, looking after the divers and managing their accidents. Now, roles are reversed. I remembered the tumult of the past few hours and strongly appreciate the deep diver's vulnerability. He inhabits an alien world. He is extraordinarily dependent on the thoughts and actions of the topside team. Suddenly I see the diver's world in new perspective.

"Joe, we're going to hold you at this depth for an hour. We'll increase the partial pressure of oxygen and send you down some food. Heinz and Pat will go and talk to the computer. At the end of an hour we should have a new schedule. Then we'll start home again."

I welcome the confidence. My pain has diminished into a tiny electric trickle of discomfort. Bev leans over and smiles.

"Joe, I'm glad you had them turn around when you did. My own knees were getting pretty bad, too. I was going to speak up, but you beat me to it. As soon as they pulled the plug on this thing, the pain started to go. Not a minute too soon."

Dinner arrives. It comes with the iron clanking of the medical hatch opening and closing. Gas is vented into the small lock. When the pressure is equal to the main chamber, the hatch is swung open. Our food is still warm. It has been artfully curved to fit the small lock. Some blessed soul has stashed in two demi-tasses of wine.

The hours creep like a stagnant, weedy river. Sometime tomorrow my rubber knees will carry me out through the main hatch. The room will seem surprisingly large after the chamber's confinement.

Ahead of me lies the long road up through the shallow depths. I will spend many hours on an oro-nasal mask breathing system. It will abraid my skin. My face will come to feel like sandpaper. My lungs will burn from hundreds of liters of dry oxygen breathed through the mask. Throughout will be the threat of another ambush of pain.

But now I am content. The finger of wine rides well, and a blanket of sleep reaches for my brain stem. Tomorrow, pressure will release its grasp. I glance up at the closed circuit television eyeing me like big brother from one of the viewports. It winks back.

Lockout submersible *Deep Diver*
in pre-dive checkout of launch and recovery,
aboard *John S. Cabot*. Hills of Newfoundland
are in the background.

Descent Six
Grand Banks
1967

The first call comes late at night. A telephone message from Washington. It is my old friend Ed Wardwell.

"They've lost a classified object on the Grand Banks, somewhere south of Newfoundland. A few days ago they located it and put two Navy divers down to fasten a lift line. I hate to tell you this, but they lost both men. Apparently they got into serious trouble on the bottom and were killed during decompression. Ocean Systems has been asked to help out. We're putting a team together around *Deep Diver*. Can you be in Boston tomorrow?"

I catch the first plane. I am met by a team of old hands familiar with the sub. At dockside things are secured and stowed on the *State Express*, a fast offshore boat. Soon we are steaming north-east, bound for the rocky south shore of Newfoundland. *Deep Diver* is firmly cradled on the stern of the ship.

On the way north the weather is surprisingly serene. The Gulf Stream carries us on marble waters past Maine and Nova Scotia. Our track takes us east of Sable Island, haunted home of countless shipwrecks.

As we close on Newfoundland the weather turns contemptuous. An immense shoulder of fog rolls in to wreath the boat. We grope our way northward seeking the remote mid-sea rendezvous. The air turns sick and cold.

One night the wind blows in harsh from the north. The sea becomes malignant, and beats the ship with rigid roaring waves. Sleek barrel combers break across her swerving bow. Salty flesh of the ugly widow maker.

We limp into the rendezvous. Two ships are waiting—the *John Cabot* and the *Utina*. Slinking somewhere behind the shapeless horizon is a covey of Russian trawlers. They are well hidden, and outnumber us. They listen with submerged electronic ears.

Our mission is classified. We are to locate and recover a large cable plow. It is used to bury submarine cables under the sea floor. It was lost in 400 feet of water when its chain bridle snapped. There is talk aboard ship that the plow was burying a strategic communications cable used for anti-submarine warfare.

Our first efforts are impotent. Dives are cancelled or postponed because of weather. On one excursion we finally locate the plow and its pinger. After two hours of cold bottom searching, we are homing in.

Just as we are bearing down, a call comes from the surface. "Return at once. A thick fog is rolling in."

We barely make it back on board the *State Express*. The waves smash the sub into the stern of the ship. Our left bow plane shatters like glass.

We move the sub and her cradle to the *John Cabot*. Her forty-ton crane is an improvement, but the thirty-foot lift from the sea to the deck is terrifying. Time leaks away. Only a few hours remain for one more try.

A September wind cuts across the sea, but I cannot feel its bite. The ocean is gaunt and cold, but I cannot feel its chill. I am a prisoner, suspended in a steel sphere, high over a wild and rolling plain. Below me the ocean is bruised and dimpled steel. Its naked waters flee before the bleached teeth of the wind. It is evening on the Grand Banks, south of Newfoundland.

At my shoulder is Roger Cook, pilot of our small research submarine. *Deep Diver* dangles like a loaded cannon at the end of a long cable. All twenty-two feet of *Deep Diver*'s curves and straight lines are firmly held in a rubber-steel cradle. Her litany is: Weight: 8-1/4 tons. Beam: 5 feet. Payload: 2,000 pounds. Operating Depth: 1350 feet. Lock-out depth: 1250 feet.

Like any lovely lady, her statistics are impressive; but they only hint at her total performance. Her ability is best realized in the blue water-bed of the sea.

Roger sits, tensed, waiting for the next move from the launching crane. We have just been lifted from the deck of a large ship and are about to be lowered into the sea. My world-view is through six small windows in the bow of the sub. I move closer to the icy plexiglass.

The sea spins and rotates. The ship swings into sight and then disappears. It looks like a dark lighted building lying uneasily on its side. Overhead, the sky is a rough asphalt blanket flowing over the waves.

"Get set, guys. We're really going to swing on this one."

It is Roger. His words are like chopped iron. A huge roller lifts the ship. The crane's boom tilts skyward. Our eight-ton submarine arcs forward.

"Jesus. We just missed the hull. They better get us out of the sky and into the water. This thing is not designed for flying."

The submarine shudders. The cable above us is being paid out; a few feet at a time. Thirty more to go before we hit the water.

I am now at eye-level with the sub's support cradle. Its skeletal frame lies on the ship's helicopter pad. Four men stare at us from its outer edge. Our surface support team. Their faces have the grim look of people at the scene of an accident. Their clenched fists hold limp white nylon lines used to guide us across the deck. Now we are free, except for the single serpent grasp of the overhead cable.

A leprous wave sweeps in under the sub. Its briny fingers fail to reach the steel.

"Stand by. I think they're going to drop us in on the next one."

Roger's voice is forced determination. He hates being airborne. Both he and his sub are at home only in the sea.

Our timing is off. The next wave rocks the sub with a sullen arctic slap. We roll madly in its foamy wake.

"Slack the damn cable. Let's get free of this bloody surface."

The words fly out from the diving compartment behind us. They belong to Denny Breese. He also feels safer when the sub is well below the anger of the waves.

"It's okay, Denny. The Navy divers are just climbing on the sub. They'll unshackle us in a minute."

Roger strains forward in the conning tower. Its eight ports give him a panoramic view of the turbulence. He sees two black figures scramble up onto the sub. Behind them is their rubber dinghy.

The weight of the two divers tilts the sub. They move quickly to the mid-section just behind the conning tower. There is the clatter of chains and cable. Our pelican hook is off. We are free of the lift line. The divers plunge into the waves and fight to gain their rubber boat.

The sea swirls around the ports. We grind and heave. I hear the sudden sound of steel on steel. The sea has pushed us against the big ship.

Roger powers the main motor. A propeller whirrs in our tail section. *Deep Diver* responds sharply. We lunge forward and clear the ship.

Our bow points into the wind. The rolling stops. We now pitch into the trough of each wave. At least we are out of the air and in the sea. A confrontation under control. We begin to unwind.

The lights of the big ship glow wearily through the falling darkness. She is the *John Cabot*, a Canadian Coast Guard cable layer. The command vessel of our three ship flotilla, 150 miles south of Argentia, Newfoundland.

Deep Diver slows her forward motion. Roger moves positively, checking the trim in the forward trim tanks and testing full series power.

"We're ready to dive. All set in the rear compartment?"

"Aye, ready. Any time, Rog."

I look back through the short tunnel which connects the two compartments. Denny Breese and Vince Taylor adjust their diving gear. Their breathing regulators glisten chrome. Their suits look like faded green parchment. At four hundred feet the water outside has a polar heart. Thirty degrees. Even the two sets of wool underwear won't keep them warm for long.

Vince Taylor is a commercial diver who plies his trade in the Gulf of Mexico. For years he has worked deep beneath the dangerous steel frames of offshore drilling rigs. He does not like cold water. The Gulf sea is almost twice as warm as the liquid frost which swirls at my view-port.

"When we find that big mother, you can be sure that ol' Vince is gonna be fast on his feet. Open the hatch and scat, I'm gone. I'll sling that big iron harrow faster'n you can say New Orleans."

"Great," says Roger. "I'll feel a damn sight better when we get back on deck. The wind is makin' up, and our recovery is going to be much worse than the launch."

The bat-wings of isolation beat inside my stomach. The sub slides forward. The sea closes in.

Within seconds we are well beneath the waves. The heaving ceases. It is like slipping a canoe into a calm pool after a wild ride through the rapids.

Roger flicks on the exterior lights. Black eddies disappear. We are encompassed by a halo of pale green fire.

There is nothing to see; nothing to judge things by. The water drifts off into velvet infinity.

Then they come. Tiny white satellites floating past my window. Life. Alive in the sea. Hundreds of minute organisms in aimless ocean drift. Pin pricks of pulsing jelly against black obscurity.

Roger speaks to us.

"I've got a compass heading on the Navy ship. I'm going to surface and power over to her. Let's see if we can pick up a visual on the grapple wire."

Roger points the submarine due north. He is heading toward the *Utina*, a small U.S. Navy support ship. One hour ago, her crew finally hooked a grapple into the lost object. Hanging from her stern is a long thin wire connected to the plow. The wire is not strong enough for lifting, but will save hours of search. We can follow it to the bottom and then drop the divers out. They will fasten a more substantial salvage cable. Strength and security of the cable is important—the plow is bigger than two bulldozers.

This is our last dive. Each day the weather has worsened with the imminent approach of winter. Earlier attempts have been plagued by the black weather and our inability to locate the plow. This time we are confident. All we have to do is follow the cable to the bottom.

The submarine fills with the clatter-whirr of hydraulic and electrical sounds. Everything is dark, except for the crimson glow coming off instrument panels. A submerged aurora. Roger moves deftly, sure of each practiced action.

Behind me I hear Denny and Vince talking softly. They are rehearsing the steps each will take when Vince exits the sub.

"I've got the cable in sight. Can you see it, Joe?"

It lies straight ahead. A dull black thread, hanging deep and bowed in the sea.

"She doesn't look too tight, Rog. Can you get *Utina* to take up the slack? It'll make it a lot easier for us to follow it down if the cable is straighter."

"Okay. I'll ask."

Roger talks quietly to the surface crew over *Deep Diver*'s intercom. I watch the cable begin to tighten.

The wire has a certain meanness to it as it sways slowly in and out of our lights. My brain shivers at the thought of entanglement.

Roger hovers the sub's bow towards the wire. He carefully adds water ballast and we begin to drift downward.

It is hard work for Roger. The sub wants to spiral away from the swaying cable. Calmly he guides *Deep Diver* back on target.

The high, black coffin lid of the sea closes in high above us. Our depth gauge rises. It's a long way down. About the same as the Washington Monument or Ottawa's Peace Tower.

Seconds pass like water dripping off ice. The wire is held firmly in the forecourt of light just ahead of *Deep Diver*. Behind it I see a ripple flash of silver. Fish. Big ones. Hundreds of them. Swerving in to see us pass.

"Three hundred and fifty feet. Fifty feet to go."

Roger's voice is firm and clear. The prize is almost in his grasp.

Then, like a whip-crack, the cable disappears.

"Damn. Where in hell did it go?" Roger curses. He holds the sub steady.

The intercom crackles with wet static.

"Sorry, you guys. The *Utina* just slipped power. She can't keep the cable tight."

Roger's response is instant. He reverses the main motor and backs us away. The cable is loose somewhere below. When it tightens, it could draw into the sub.

I look up at Roger.

"Rog, why don't we forget that rampant water snake. We're almost on the bottom. If we drop straight down we should be awfully close."

Roger hesitates. He's not anxious to lose the first direct contact we've had with the plow.

"I'm not happy about it, but let's try. Sure don't want to get snagged in that thing."

He picks up the intercom to inform topside of the new plan. I reach over and turn on the carbon dioxide scrubber. Its breath is warm and comforting in the dark interior. I check the oxygen flow meter. The precious gas trickles in at two litres a minute. About as fast as we use it up. I return to the viewport.

Below us is desperate soil. Stark clay and rubble sand, brown stones and flat rocks. Stubbed weed. All drawn into the two-dimensional light of our artificial moon.

"Bottom in sight. Ten feet below."

I am overwhelmed. The Grand Banks. The floor of one of the world's most fabled fisheries. Living magnet for North America's early explorers. In the hidden distance I see another flash of silver.

We touch down softly on the dark turf. Steel thunks against stone. Another distant animal glitter.

My mind flicks back to a huge squid lying preserved at the university in Newfoundland. Almost forty feet long in a glass-topped case. Suction cups as big as saucers. Ugliness in alcohol.

"What the hell is that?" It is Roger. His voice is weathered.

The cloud grows bigger. A silent brown boil. Alive. Reaching out. It cares for no man and no thing.

"My God. It's the cable."

A malevolent whip flashes out of the cloud and disappears. One end of its black snare form leads deep into the roil.

"Jesus. The cable is hanging in the mud beside the plow and flaying the hell out of things. We better get out of here."

A motor groans. The sub slips into reverse. I see where the sea floor has been whipped by the feverish wire. Scalpel thin grass points up between indentations. Broken greenness.

We pause in the leer of our own lights. Everything is silent and grim. Minds click like camera shutters over possibilities. Denny speaks up from the diving chamber. As always, he is optimistic.

"Why not circle around and come at the plow from the opposite side? That's probably our best chance of avoiding Super Snake."

"Okay, Denny. I'll work us slowly around. Let's see if we can't come in 180 degrees from where we are."

We edge away from the wrath of the snare. The cloud fades behind us.

Roger assumes the loose tenseness of a fighter. He guides us around the circle as if manoeuvering an opponent to the opposite side of the ring.

I reach for the head set of our sonar receiver. It is tuned to pick up the sounds of the plow's pinger. I hear a high, vague tone and adjust for direction. We complete the half circle and squat down in the mud.

"We better find her on this run. I'm starting to get cold."

It is Vince. His Louisiana blood is beginning to chill. So is mine. The hull seems coated with hoarfrost. My wools are almost defeated. We have been down over two hours.

"This is it, Vince. This is the run. We're going to put you two guys right next to that bloody thing."

Roger lifts the bow and we slide foward. Gravel grates on the battery pod. It is an abject sound, like chalk on a blackboard. We begin to close on the plow.

The sounds from the pinger are useless. High-pitched baby shrieks. They come in from everywhere. At such tight range the pinger is omini-directional and unreliable. I put away the headset and hunker next to the viewport.

We move slowly forward through drenched hollows. A sleepy seeming fish cast in metal. Our eyes vent anxiety through the markless wilderness. Only blotches of darkness return. Black holes in the sea.

Then I see it again. The hueless brownish cloud. It now looks like rampant thunderhead bridged with high anvil.

"Damn. We've missed again. We've gone right by. We're back where we started."

I curse the sea. I curse the cable. Mud swirls near the viewport.

"We better just stop and hold fire for a few minutes. I want to figure things out." Roger sighs the words.

The submarine squats just out of range of the cloud. Each of us pauses in shocked thought.

I stare at the enormous horizon. Our rime-ringed light runs out to the wall of cloud and halts. Behind it lurks the tormented cable.

I worry about our next decision. We are tired and angry. Frustrated. A time when courage can melt from bravery into foolhardiness.

The intercom crackles. "*Deep Diver. Deep Diver.* Request that you surface immediately. Surface immediately. Winds have increased to 30 knots."

There is no mistaking the urgent tone. We must leave right now. The decision is no longer ours.

"Well, it's been nice masquerading as a diver for these past few hours. Since I don't get to swim, no one will know the extent of my talents. Remember the oath: 'We dive for five. . . !' "

Denny rambles on, prying the lid off our tension. I look into the ocean and see the twist of wire inside the cloud.

Roger backs down the sub and begins to pump ballast. We lift off the sea floor. Slowly we begin to climb up the dark staircase.

Fifty feet off the bottom the water is much clearer. White dots of tissue again float by our lights. A snow storm. A blizzard of drifting plankton. Grass of the sea. Food for the countless millions of fish that swim over these silent banks.

I see a touch of silver. It becomes a wall. Television sharp. Etched on black. Cod. King of this sea.

We float up in an outermost ocean. Hundreds of square miles of continental shelf stretch beneath us. Ours has been the first glimpse of this new frontier. We saw something less than an acre.

Others will soon come. Drilling ships will probe here for oil and divers will walk this airless earth.

Many have gone before, but always with empty eyes. Not far from us, away down the continental slope, lies the huge broken hull of the Titanic. The queen of shipwrecks.

The soaring ceases. Roger holds us at forty feet, just below the down-reach of turbulance. Life continues to race toward our lights. Swarms of white pulsing dots. Insects of the sea.

"*Deep Diver. Deep Diver.* We think we see your lights aft of our stern. You are three hundred feet away and clear of surface vessels. Come up when ready."

From high on the big ship our submarine lights must look like a glowing emerald pool. All hands are at the rail to watch our recovery.

"Urge extreme caution when you come alongside. Winds are at thirty knots and gusting to fifty."

An unwhispered dread creeps through the sub. I clear the smoke of fear from my brain.

"No sense waiting, Rog. Let's give her hell."

I put my back to the hull and brace my legs. Air races into the ballast tanks. We drive towards the surface.

Deep Diver breaks through like a sperm whale breaching. Ice-blink waves tear at the conning tower. We roll wantonly in high breakers.

The sea is pandemonium. Waves break like bones around the steel. All viewports are shrouded in a whirlblast of deep-sea rain. We roll and pitch. From somewhere in the metal spheres I hear the moan of protest.

Roger powers hard toward the dying lantern lights of the big ship. In the shadow loom of its hull he throttles back and coasts.

I hear a low volley of thunder. Waves are breaking against the big ship's hull. We move towards the lee.

Roger works feverishly. He kicks the stern around. Too late. Metal grinds on metal. A short tearing scream.

The submarine jumps ahead. Heavy breathing utters from the dark interior. The smoke of fear returns.

We slip away from the hurl of the next wave. Two divers appear from under a lamb white wave. Chains clatter overhead. I hear the deep click of a trap shutting. The pelican hook is set. We are fettered to the lift wire.

Both divers are swept back toward their raft. The sea feels nothing.

We are lifted free of the seething tangle. For a second there is no movement and then the slow pendulum begins. At first the new motion is seductive, almost comforting. Like new love sleeping on a white shoulder.

I hear a distant tearing sound, like a backstay screaming in a storm. It is the wind on our lift cable.

The pendulum swing increases. The ship comes near, and then falls away. We move slowly upward.

Our swings parallel the ship. Bow to stern. The black steel plates are perilously close.

Transfixed to the porthole, I helplessly picture metal mating to metal. We arch in tight to the hull. One of the ship's portholes is directly opposite. We brush so close that I see a sailor sitting in his room. In a warm light, reading.

The strong arm of the crane shivers us up. Now we are level with the helicopter pad. The crane rotates us towards our cradle. Figures rush in below. Phantoms in the rain. We drip and spray them with wind-torn water.

The submarine becomes an enraged bull on mad Pamplona's streets. The phantoms rush at us trying to secure steadying lines. Their timing is hopeless.

We pause for an instant. There is the rough click of iron casts and three hooks are secure. White nylon lines go taut in tense hands.

The deck is suddenly alive with men. The ship's crew appears from nowhere. Strength in numbers. Five on a line but still towed across the deck by the bull's wild rampage.

The cradle is close, tantalizing. And then we swing. Dangerously close to the heads of the matadors. Nine tons of frothing, sweating anger.

We can do nothing inside. We have lost control. We brace ourselves and watch. Trip hammers pound in our chests.

The interior lights up as if in a sun gun. A wild swing has put us almost into a deck shed, with its flourescent ceiling.

We shudder into a narrowing lane. We crash into the cradle. Inscrutable silence. Wind-sigh off metal.

I am the last man out. Slowly, I climb down from the conning tower. The three others stand on the deck staring back at the sub. Their eyes are as empty as broken windows.

Descent Seven
Seven Hundred Feet
1968

It is March, 1968. Edwin Link's ship *Sea Diver* has made an easy crossing of the Gulf Stream. The afternoon sun silhouettes two lone gulls diving for scraps in the loose froth of our wake. The engines slow to half speed. From my position in the crow's nest I see the low green outline of Great Stirrup Cay and two smaller islands to the south. To the east is the pale finger intrusion of a lighthouse.

The high sun gives bold relief to the coral reefs ahead. Waves break like soft cotton and reveal the rusty teeth of staghorn coral.

The engines slow to full stop. *Sea Diver* continues coasting. Her bow wave falls away and the warm froth of her wake disappears. The signal is given to lower the anchor. I hear a discordant tympany of metal against metal. The anchor chain runs hard over bow chocks. The sound stops. We are anchored in sixty feet of water. Our berth is secure.

I scramble down from my perch high above the ship and join the others below decks for dinner. John, our Greek cook, has steaks tonight, and the aroma surrounds the ship.

I have been to the Bahamas many times, but this may be the most important trip I will ever make. In the next few days we'll attempt the world's deepest working dive. One man made a brief pause at greater depths, but this will be the first time that divers have actually carried out meaningful work. At dinner Ed Link summarizes our position to Jeb Gholson. Jeb has just arrived to make a film on the project.

"We should be ready. For the past month we have made forty dives throughout the islands, slowly working up to 'the big one'. Our expedition began in Nassau. For several days we made shallow excursions

Deep Diver being launched
from stern of Link's ship, *Sea Diver*,
just prior to the 700-foot dive off
Great Stirrup Cay in the Bahamas.

south of New Providence. Then we steamed down along the east edge of the Tongue of the Ocean. Our next long stop was at Gould Cay, where we made a rather extensive series of deeper dives. Near Fresh Creek on Andros we did some two- and three-hundred-footers. The submarine *Deep Diver* is working well. All the lock-out dives have gone smoothly.

"The last excursions brought us to a high peak. Several times the submarine was positioned on a narrow ledge almost two hundred feet down. The divers compressed, dove out, and studied the 'Wall" below the sub. It's called the wall because it's the sheer side of a submarine canyon six thousand feet deep. It was a splendid opportunity for the divers, several of whom were biologists. In a short distance down the wall they had access to a constellation of sea floor life. Dr. Walter Stark found several new species."

My mind returns to the "wall." It brings memories embracing both fear and achievement.

When divers lock-out, the submarine is ballasted heavy to prevent it from drifting away from the divers. The ledges on the wall were extremely narrow. I could see the steep sheer of the cliff above and the empty abyss below. A single error could have slipped the sub off the ledge and down six thousand feet—well beyond collapse depth. Fortunately all went well, and some remarkable discoveries were made.

After dinner we begin the final systems check for tomorrow's dive. The submarine team gathers at the stern of the ship to run down the check list. Working beside me is Denny Breese, diver, mechanical/electronic specialist, and ex-nuclear submarine technician. Like the others, Denny does many jobs and does them well. Checking the bow thruster is Roger Cook, submarine pilot, equipment genius, and an extraordinary diver. Roger and Denny have opted for the most difficult and dangerous task—to compress themselves and swim out from the sub at maximum depth.

The third key member of the support team is George Bezak. George's main concern is the intricate electronic system. Short and soft spoken, he too is a superb pilot and diver. Breese, Cook and Bezak—a splendid team.

Later, all four of us again gather at *Sea Diver*'s stern. Small waves slap softly against her heavy steel sides. A brilliant pearl moon washes the lower sky. Together we make the final inspection of the vehicle that will carry us into tomorrow. Even at this late hour there is time to anticipate and correct problems.

Walking past her stark yellow shape I marvel at the pinnacle of technology which she represents. She is the only submarine able to make a 360-degree turn within her own axis. She is the first to combine the speed and mobility of a submarine with the work potential of a diving bell. She is a mobile base for divers. The most flexible undersea system.

The rising moon glints off the conning tower viewports. There are enough of them to give the pilot 360-degree visibility. I watch George Bezak slip below the round plump hull to recharge the bank of batteries in their black jettisonable pod.

For a moment I stop to look at the prominent flash coming from the Great Sirrup lighthouse. In 1964 we were anchored in almost this same location for the deep saturation dive. Almost four years ago. The busiest years of my life. Years, that for all their activity, have melted away.

The black warp of the shore stares back through the night. It is time for sleep. As I turn to go below to my bunk the breaking surf echoes warning—be careful, take care.

A gentle lift sensation and we are free of the stern cradle. Within seconds we are awash and held by the sea. Crystal clear water sweeps over the viewports and sunlight dances back and forth above me. From my co-pilot position I look out the lower viewports into blue shimmering curtains which disappear into the unknown water.

The electronic system shirrs into action and the sub begins a soft but insistent forward motion. *Deep Diver* has slipped her tether and is free. The dive has begun. I start my stop-watch: 1:47 p.m. (1347).

1347

We begin pumping water into *Deep Diver*'s trim tanks. This will stabilize us while submerged.

1350

George Bezak, the pilot for this dive, works deftly on my left side. He drops the sub just below the surface to check the trim and communications system. It is a cool and welcome relief. The hot sun streaming through the upper ports is too warm. Roger and Denny run over compression procedures in the diving chamber behind us. Their quiet voices carry through the open hatch.

1400

We surface momentarily to take final bearings. John Barringer, topside co-ordinator, informs us that a fibreglass faring has worked loose. The sub tilts slightly, as a diver climbs up near the conning tower. While on board he removes the faring, and inspects a nearby oxygen pressure line. He confirms its integrity. He also reports three large sharks on our stern and then quickly scrambles back up on *Sea Diver*.

1421
We begin to descend.

1423
I still see the moving mirror surface overhead. The sub angles down as if on a gentle slope. Our depth is 25 feet. Heading: 350 degrees.

1429
80 feet down. Two sharks follow our stern as if attached by invisible wires.

1450
Depth 380 feet. Below us the blue has given way to a granite darkness which rejects my vision. However, I still see light overhead. We hold for a communications check. Topside informs us that we are coming through "loud and clear." Although the radio voice is strong, I feel the thinning of our physical links to the surface. It seems we are descending into the centre of the earth. The sub continues to drop. More tons of ocean water close in overhead.

1456
500 feet. No bottom in sight. None of us expect to see it, but all eyes are earnestly probing. Seeing the bottom will reduce apprehension and the helpless feeling of being suspended over infinity. Darkness grows. The lights on the gauges and panel beside me become more prominent. Our voices hold us tight together within the cramped confines of the sub.

1500
600 feet. Continuing slow descent. Our search for the bottom is still in vain. Complete blackness surrounds us. Like being 600 miles out in space.

To the novice the noises of the submarine would be completely unsettling: a loud cacophony of whirrs, clicks, and the occasional slurr of hydraulics. For us, the sounds are comforting; but any change in pitch or tone causes a shiver of alarm.

1514
680 feet. I sight the bottom about ten to fifteen feet below. It is hard to tell how far away it is, for there is no scale of reference.

1516
George keeps the sub hovering near the bottom. A one-knot current

angles in from 300 degrees. It moves the sub with its impressive sway. Likely a backwash from the nearby Gulf Stream. With the lights off I can see about thirty feet ahead. Amazing: incredible, that enough light passes through 700 feet of water to illuminate the dim outlines. A tribute to the human eye and the clarity of Bahamian waters.

1518
George alters course to 270 degrees. He's sighted the downslope, and will slip down the flat sand bottom until we reach lock-out depth. The floor is like a desert. Sand ripples glide geometrically below us.

1521
693 feet. Progress is slow. We cautiously seek the right depth. The bottom has no rocks or coral, and only a scattering of small plant life. A large fish occasionally flashes into our light beams to inspect the new predator.

1525
700 . . . 700 feet. I recall looking down the same distance from an office building and being amazed at how small things look. I feel over-whelmed at the significance of our being here.

1530
We agree on the site. George begins to flood the ballast tanks. The whirring pumps vibrate. The sub settles squarely onto the hard sand. Thunk.

1534
Both tanks are flooded. We sit hundreds of pounds heavier on the sea floor. A sudden picture of not enough ballast flickers across my mind. Roger would step out and move away from *Deep Diver* . . . turn around . . . and find the sub drifting up and away from him.

1537
Our words are slow and measured. Adrenalin begins its tide flow in our veins. We rest on a slight slope with the high side at 240 degrees.

A sudden strong current sweeps across the slope and drives *Deep Diver* around like a weather vane. The divers are too busy to notice, but George and I exchange puzzled glances. Is there too much current? Should we risk it? We agree to wait. We'll check if it's a constant flow or a one-of-a-kind eddy.

1604

The current changes direction and turns us slowly to the left. We are on the weather edge of giant spinning vortex.

1605

George calls me up into the conning tower. Its circle of windows is covered with a splash of silver fish. They swim backwards and forwards trying to get closer to our interior lights.

1609

The current has now dropped to half a knot. It is steady. A quick discussion. Agreement. We'll go ahead. The orange caution light flashes in my brain.

1610

I close the white aluminum hatch between our chambers. The divers are sealed in. In a few minutes I will not be able to reach out and touch them. An excess of metal and pressure will soon create an incredible distance between us. I will only be able to see them through the viewport.

1612

Words stop flowing. Silent thoughts and ancient practiced actions take over. We begin to pressurize. Gas roars in.

Four hearts accelerate simultaneously. High adrenalin time . . . sweaty palms and clamouring hearts. Time to carefully sequence thoughts to flow in an orderly fashion. Time to keep the brain steady.

1614

Helium-oxygen gas screams into the diving chamber. Roger and Denny have four hundred feet of pressure around them. It builds quickly.

They appear calm, although no one knows what they are thinking. They are surrounded by shrieking noise, sauna heat, and the violent rush of increasing pressure.

No voices travel over wires. The sub is a moving crescendo of sound. We are all suspended: thoughts moving, words waiting for silence. Through the viewport I check for the hinted sign of a problem. Both divers are imbedded deep within themselves.

1618

Compression stops. Silence. Gas pressure inside the chamber now equal to the water pressure outside. The hatch falls open—no longer suspended by water. A balance of pressures is achieved. The new pool under their feet laps blue testimony.

1619

Denny slips into the opening, grabs the hatch, and spins it away so it will not impede Roger's exit. He stands on the sea floor only three feet below the sub.

Roger makes final adjustments to his breathing system while I check the oxygen level in their chamber.

I talk to them and confirm their condition. It is not easy for I am confronted with the insufferable problem of two voices destroyed by helium. They talk in croaked, off-key cackles. They speak slowly and use familiar words. With intense concentration, I comprehend.

1620

Roger pushes himself forward and slips easily into the pool. His head disappears below the surface.

Through a forward viewport I watch the sea-black figure of Roger. He tracks out from the sub. He swims slowly as if trying to test the water. Each breath is carefully taken. Each inhalation is made up of gas that would be fatal on the surface. At 700 feet the increased pressure concentrates oxygen. At this depth it is ideal for his body.

Roger's passport to life is a breathing system releasing part of every exhalation into the sea. The rest remains in the breathing circuit where it is scrubbed free of carbon dioxide. It is replenished with more oxygen-helium, and returned to his lungs.

Roger moves like a slow-motion ghost among shadows. His black form drags a light umbilical tail leading back to the chamber. In the sub we hear the draw and release of his breath. Very controlled. His hand reaches down to pick up a green leaf algae from the sea floor. His eyes lift and stare into the circles of black outside the light.

The sub is frozen silence. George and I listen intensely to Roger's rhythmic breathing and occasional helium words. I am comforted by the periodic escape of gas from his breathing bags. I watch the bubbles and wonder how big they will be at the surface. Probably they will expand to the size of the sub.

For a second I allow myself to dwell on the exhilaration of the event. I am watching a man make history. A look into a window of the human spirit. This is man's deepest working dive.

1624

Denny has joined Roger in the water. He breathes from an open circuit system and has less time available. He moves softly but positively in the water. An alien ballet dancer. After the dive he will tell me,

"What first struck me was the visibility. With the lights off I could see about 50 feet. When the lights came on my vision was reduced to the yellow cone of light. Everything else went flat black."

1627
Roger heads toward the submarine and attaches his net bag to it. It is full of algae. Seconds later he is inside the chamber dripping wet, and grinning wildly. Then he disappears. He has joined Denny to take pictures of an elusive moment.

1632
Denny is now inside and calling Roger to return. Sweat beads my forehead. We are heading into a critical corner. Each minute of exposure at the end of a deep dive adds huge blocks of time to the sentence of decompression. We have agreed on thirty minutes out, but I want it to be slightly shorter to cushion any unexpected problem.

Denny continues to call Roger in. No response. I add my voice. Roger, savouring the moment, is not eager to release it.

"I think I was slightly euphoric. It wasn't narcosis, but I sure felt good. I now know why Ed White didn't want to climb back into his space capsule . . . he wanted more of that very special freedom."

1633
Logic overrides. Roger slips back through the aluminum ring separating water and gas. Now they are both inside laughing. The infectious helium giggle of success. Roger strips off his gear and helps Denny secure the hatches. They move fast. Time is running out.

1634
The external hatch is tight and the sea is barricaded out.

1635
The internal hatch is secured. This one holds high pressures inside the chamber when it is on the surface. Both hatches are critical. One to keep out the sea, one to keep in pressure. My voice begins to lose its steadiness as we begin to move into the most dangerous part of the dive—decompression. A reef of hidden dangers.

1638
We lift off the bottom and start towards the sunlight.

The sub is at 650 feet; the diving chamber at 700. I begin to release pressure on Roger and Denny.

First stop in the decompression is at 480 feet. Suddenly I am impossibly busy. My actions track along a hurried path: watch the divers, listen to voices, monitor external water pressure, control internal gas pressure, keep an eye on the stop watch. My hands and eyes move rapidly from place to place—stopping only to pick up information or move something. An open tape recorder collects my summary of events.

During this part of the dive, it is essential to compare external and internal pressure. External must be lower, otherwise gas will not flow out of the diving chamber and decompression will stop. The consequences are disaster. George helps by accelerating the sub to the surface. I coordinate opening the exhaust valve to his ascent.

1639
550 feet.

1640
450 feet. We are moving fast although it is hard to tell from inside the sub. No feeling of motion, only a slight upward tilt and more light pouring through the upper viewports. Someone is turning on a giant blue rheostat.

1642
250 feet. To keep ascent control George vents a huge swarm of metallic bubbles. They clatter like pebbles against the hull.

We rush out of the blackness outracing exploding gas. A yellow shape struggling from a darksome vault. A yellow form trying to shed tons of water, we struggle like a rigid whale. The sub disgorges an enormous silver tail of helium bubbles. Everything trembles and rushes as if to escape the maw of some unknown monster.

1649
We burst through the surface in a shower of white. Glistening water and foam is everywhere, mixed in with the welcome sun. The sub plunges forward as if to glide off a precipice, but it is only a righting movement. In seconds, we are back on the sea, caught in the trough of waves. Denny reports a big amberjack just behind his viewport. Our motion is all rock and swish-fall. The indigo seas have grown in size since we left the surface. We feel uncomfortable with the new motion.

1652

Roger and Denny are at 430 feet of pressure. Decompression continues smoothly. Then . . . crisis!

A gas leak. The pressure begins to fall in their chamber. I add gas to hold them at 430. Strength begins to ebb from my pores. I realize how critical the situation is. I ask the divers to look. Denny finds it. A tiny valve on the inner hatch. Partially open. A quick turn and he secures it.

I offer thanks it is something so simple. A jammed hatch could leak the life away from the two of them.

I note how easy it would be, with nerves so taut, to be swept away by the fierce night winds of panic.

1703

We steam slowly toward *Sea Diver*, easing across the waves towards us.

1704

Divers at 360 feet. I breathe easier. So far the journey is successful.

1705

George discovers we cannot use our bow thruster. Each time he applies power it trips a circuit-breaker. This failure makes it harder for him to position us beneath *Sea Diver*'s crane.

1709

350 feet. Both divers feeling well. Once again we slip into the stern shadow of *Sea Diver*. George deftly guides us into position. The lack of bow-thruster doesn't seem to matter. The sharks again take up their promenade behind us.

1713

Secure on deck. Cradle-lashed. We are now a part of the mother ship.

1720

Decompression moves into its slower stages. Divers are at 320 feet. The ship has weighed anchor. We are on our way home to Florida after a stop-over at Bimini.

1745

All goes well. Both divers resting and quiet. The ships engines vibrate through the hull and find me still in the forward compartment. I am alone but more comfortable. The pace has slowed. There is more time to watch and maintain pressure, observe the divers, and make entries into the log.

1749

Another black pinnacle looms. Denny. Over the intercom his soft voice describes the weird sensations that surround him.

"I have the feeling I am spinning wildly—that is everything around me is spinning. It's especially bad if I move my head. It seems better if I shut my eyes and hold my head firmly against the hull. What do you think, Joe?"

My spirit sags. We are up against one of the ugliest of diving problems—decompression vertigo. A disturbance of the inner ear and the body's balancing mechanism. Very close to the brain. What makes it so vexing is that little is known. I sit in haggard solitude and I try to conceal my dejection.

There is no question about what to do. I stop decompression and pull the full story from Denny. I watch him as he talks. His eyes move involuntarily—back and forth . . . as if he were watching a fast tennis game. The symptoms unfold.

I increase the pressure until his symptoms subside. I add oxygen to his breathing mixture. We back down to 320 feet and begin the wait for this unwholesome mountain of a problem to resolve. It does. Only partially.

I do something that I rarely do in decompression treatment. It masks the effectiveness of pressure therapy. I give Denny a drug to still the rage of his spinning. Within minutes he feels better. We wait.

The next step is critical. I always hesitate before taking it. I decide on a new schedule to bring the divers up. Our laboratory support team has provided several options. It is a much more complicated process, due to the tiny "bruise" in Denny's ear. If damaged again, it might affect his hearing and balance for the rest of his life. If really severe, it could send him into shock.

2230

Five painfully long hours have passed. Denny is feeling much better. The new and time-lengthening schedule seems to work. Both Rog and Denny drop off to sleep. The ship continues its journey into the night.

March 7th, 0110

A new day. We anchor at Bimini until dawn. Through a fogged viewport I make out a radio tower stabbing its red lights into the night. It is not quite 700 feet high. . . .

0700

Sea Diver lifts anchor and heads into the Gulf Stream. For the next eight hours we lock in another kind of combat with the sea. A strong west wind drives burly waves over the bow. I sit as long as I can in the cramped compartment. Stale air and the furry edge of sea-sickness force me to stand up in the conning tower. Deep lungfuls of fresh air help.

Tonight Roger and Denny will breathe the same air for the first time in almost two days. They have dived in the Bahamas and will surface in Florida. A strange kind of first. . . .

The wind blows stronger. I feel needle points of spray on my face. Tonight we finish. The divers will surface. Somehow it will be anticlimatic. The best part was the stolen moments of euphoria. They are hidden forever far below the distant waves.

The divers will decompress just before midnight, and as soon as they are on deck, the talk will turn to future plans for an even deeper dive—perhaps to 1,000 feet.

I will watch Ed Link's face and know his thoughts are on the improvements necessary to develop an entirely new kind of submarine. His vision has already grasped the implications of a transparent hull. For him it would be the window to a new world and even greater challenges.

Support boat approaching the
diving ship, assisting in body recovery.
Hills of Venezuela
are in the background.

Descent Eight
South American Salvage
1968

The night is flawless. The big jet comes in high from the north, winging under the pole star, a pin-point of light across the December sky.

Inside, the forty-two passengers and eight crew prepare for landing. Drinks are finished and magazines returned to the racks. Everyone moves easily into their seats, like an audience settling in to watch a pleasant movie.

The lights of Caracas are ahead, low and to the south. They wink brightly, reflecting Orions Belt in the milky way far overhead. Behind the city, cavernous black hills rise to meet the night.

The jet is alone. No other aircraft shares the soundless velvet sky.

The sea too is empty. A lone fishing boat makes its way west over the gentle swells. Its skipper gazes shoreward, eager for home.

The jet slides into its final descent. Its four engines soften their collective scream. Faces crane at windows, anxious for the first sight of land.

The crew begins the complex orchestration of final approach. The main runway at Maiquetia is less than thirty miles away. Air brakes ease on and landing wheels fold down. The plane moves into the softer air above the sea. Less than a minute remains before touchdown.

Suddenly, the plane loses altitude. A wild, hysterical shrill overwhelms the whine of the jets. A nightmare begins.

At five hundred knots, water is like concrete. A wingtip catches. The huge jet begins to cartwheel. The fisherman, glancing up, sees a long yellow crayon of light; then a series of white-fire explosions.

Aluminum alloy turns into tinsel. An exploding Christmas tree of metal flares across the ocean's surface. The sea becomes electric. The fisherman prays.

Wings, engines, and tail tear away. The fusilage is the last to lose its form. Its round shape holds it together for a few seconds longer. Then it breaks open like a child's tossed toy.

The long path of wreckage floats for some time and then begins to drift into the sea's dark arms. Silence brings a touch of mercy.

The ocean closes over its wound. Eleven miles north of Cabo Blanco the charred mosaic falls slowly on a solitary plateau; a soundless seamount sixty fathoms below. All around, the sea is hundreds of feet deep. Pieces of the aircraft drop gently onto the soft sand. The nearby waters come alive with a new sound. It is the fin-whisper of animals moving through darkness. The sharks are stirring.

Willi Myers is making the first dive.

I watch him dress into his heavy gear on the moving deck. Our vessel is the *Rescue*, an old bear of a ship with a large work area on her stern. She is inelegant, but serviceable. Her black boom swings like a lazy paw overhead.

The *Rescue* is from Jamaica. Her native crew has taken two days to work her down across the Caribbean to the crash site. She is under charter to support our diving and salvage operations.

Willi carefully dresses into his heavy canvas suit. On the deck beside him, "Big John" MacLaughlin works on Willi's helmet. A modern version of the classical "hard hat," it glistens white and copper metal in the sun.

Sixty fathoms is a depth usually reserved for diving bells. However, *Rescue* and her crew are untrained and not equipped to handle our large and complex systems. Instead, a small diving stage and classical "heavy gear" will be used. The three hundred and sixty feet below Willi is the outer safety limit for his equipment. He moves slowly.

"Take your time, Mr. Myers. Pretend you are suiting up for a black tie evening in Nassau."

It is the rough cigar voice of Bob Kutzleb. Bob is in charge of the team. Both he and Willi have been working at the crash site for over two weeks. The dives are the last step in a long and painstaking effort.

The general area of a downed aircraft is easy to find. The work which follows is much more difficult. An accurate map of important pieces of wreckage must be made. The search is for the most critical pieces, such as engines, and instrument panels. This is the only way the cause of the crash can be pieced together.

The location map off Cabo Blanco took time to make. Two beacon stations were established on the high green scarp of the shore. Radio signals from each were received on *Rescue* and interpreted as intersecting signals by a Cubic Navigation System.

The electronic results form a highly accurate grid that give the searchers an exact location between the stations. Once a piece of wreckage is plotted on a large chart, *Rescue* can return to within ten feet.

Kutzleb made his initial survey of the crash site by towing a side scan sonar behind *Rescue*. This tethered silver fish sends out and receives sonic pulses. As the sonar flys over the bottom it detects profiles of wreckage standing above the seafloor. The information is relayed to a read-out graph on the ship. It shows that the big jet was strewn along a three-thousand-foot pathway.

Kutzleb later used closed-circuit television to scrutinize individual pieces of aircraft. He and Willi lowered the camera and its rotating head over the side in a protective cage. As the camera hung over the bottom, engineers examined each piece and recorded meaningful views on tape. Then the divers were called. The salvage team would recover only the most essential components.

Willi is almost ready. His standby diver, Jim Thompson, sits fully dressed beside him. If something happens to Willi, Jim will be in the water in seconds. I am on the scene to provide medical support for the deep dives. Anxiety shares my watch. The job is the second deepest salvage job ever undertaken.

Willi shuffles casually across the deck towards the lee rail. His heavy metal boots clatter easily on the sheet steel. He turns to face us.

"Okay, topside, I'm ready as I'll ever be. It's hot in here and time to get into the water."

The sound of Willi's voice is crisp as it flows from his helmet, out of the radio speaker, and across the deck. The ship's crew moves in closer to watch.

At 5:40 p.m. Willi jumps into the water. He has decided not to use a descent stage. Instead, he will drop beside a white nylon down-line. I start my stop watch.

"That's better. Buoyancy sure feels good. So does the cool. Like a soft, blue bed."

Willi slips down to thirty-five feet and pauses. His hands circle the white line which plummets to infinity and the bottom. He checks his gear. His breathing is easy and relaxed.

"All set, topside. Going on down."

Willi's helmet timbres his voice. It's hollow. Men shuffle in closer to the radio phone. I glance at my stopwatch. Forty seconds into the dive.

At 150 feet, Bill Gianotti, the dive supervisor, switches Willi's breathing mixture from air to oxygen-helium. The new gas driving down into his helmet gives a high singing pitch.

On the surface, two men, called tenders, furiously pay out hoses and lines. One is Willi's gas, strength, and communications bundle. The other is the body-recovery line.

Willi moves rapidly. In sixty seconds he is at 170 feet. In two minutes he is almost on the bottom. He slows his descent as he approaches the wreckage. He pauses and hangs in the clear water over the plateau. He overlooks a field of torn and shapeless metal.

Willi draws his hose-bundle closer. Neither the bundle nor his rubber suit can tolerate razored aluminium.

"On the bottom. Looks like about a hundred feet of visibility."

Willi emerges on the closed circuit screen. He moves in slow motion, like a beached astronaut. Sand rolls under his reluctant feet as he plods across the dark windless veldt.

"I've got company," Willi laughs. "He weighs about four hundred pounds, and is built like Big John."

A giant grouper bulks across the screen. We cannot see his eyes, only his blurry scales and his cloud shadow frame. His tail is as big as a snow shovel. He is in tandem with a grey snapper. A smaller fish, but still weighing almost as much as Willi.

"Sure as hell isn't lonely down here. These two guys stick to me like Times Square pan-handlers."

Willi and his retinue disappear offscreen. Our only contact with him is his slow and controlled breathing. I check my stopwatch. Thirteen breaths a minute.

I pick up the intercom.

"You're six minutes into the dive. Topside gas pressures look good. Weather still like an English springtime."

"Okay, Joe. I've done a quick survey. Almost no current on the bottom. But it's running about two knots halfway down the water column. Really bends my hose. May foul up our lines. I'll only be a few more minutes."

I look to Bill Gianotti. He gives me the okay sign.

"Roger. Take your time, Willi. If possible, we'd like to bring you home on the fifteen minutes schedule."

Even fifteen minutes at 360 feet will be time-consuming. Not on the bottom, but on the way home. Decompression will take five and a half hours.

Willi is working at something. His breathing rate increases to twenty breaths per minute. I note this in the log. I'll caution him if it goes higher.

We all wait, listening. The diver in the night far below us is so damn remote. The surrounding naked sea is summer warm. A late afternoon sun washes over three gulls quietly working the swells. Water clinks against the ship like wind-blown glass.

108

Willi treks wearily back across the television screen. His boots are ankle deep in sand. He is muscle and sinew, warm behind canvas. He is sun-tanned face hidden in a helmet.

Willi is followed. An ugly, soft-pillow form. White and inert. Armless. The body of a man.

Thirteen minutes into the dive. Time to get ready for decompression if we are to make the fifteen minute schedule. The deck team moves confidently.

"Okay, topside, take me up."

The tenders begin to lift at eighteen minutes. Too late. I shift Willi over to the twenty-minute schedule. A few seconds later his easy breathing is interrupted.

"Hold everything. I'm fouled on something."

Action ceases. We all stand, poised. Impotent. Waiting for Willi's next words.

"Think I'm caught in the south TV steadying wire. I'm jammed. Can't move a damn inch up or down."

The depth gauge on the control box reads three hundred feet. Sweat runs cool rivers down my back. My skin itches.

"Hold everything, Willi. We'll slack the TV wire and see if that helps." Bill Gianotti talks into the phone and then slips away.

Seconds click by. Voices and movement are heard from the bow. Willi waits, breathing easily.

We hear the sound of light metal scraping. Something brushes against Willi's helmet.

"Okay, guys. Thanks. I've got it now. That goddamn wire caught in my hose and even had a finger on my helmet. It's thin and hard to see. I'm free and clear now. Take me up."

The tenders gather in the slack. I add the time lost at three hundred feet and shift to a new schedule. Willi moves up to the next stop. Together we climb the stopwatch-staircase called decompression.

At fifty feet he must spend fourteen minutes. At forty, an additional fourteen. He stands on the diving stage, quietly paying time penalties. The crew prepares to move the ship to the next salvage site.

Minutes pass. The sea's surface corrugates with new winds. At the stops near the surface, Willi's decompression becomes difficult. Waves gather and give the ship a loose and easy roll. Our concern mounts, for all motion is felt down on the diving stage.

More minutes pass. The sea picks up.

"I know there's not a hell of a lot you can do," Willi says, "but I want you to know that I'm now a human yo-yo. I'd be happy to trade places with any of you guys sunning yourselves on deck. Hey, Big John, how about coming down here and taking my decompression for me?"

"Okay, Superman. Glad to. Only next time we're in a bar I get to do all your drinking for you."

We take turns talking to Willi. It eases his boredom and forces him to inhale more deeply. Deeper breathing decreases his chances of decompression sickness; almost all the gas in his tissues is washed out via his lungs.

We wait, watching the sea slowly build. In a few minutes Willi will go through a phase of the dive known as surface decompression. He will be lifted from forty feet directly to the deck. His helmet will be taken off, and he will climb into a deck chamber. He will be immediately recompressed to forty feet and begin breathing oxygen. The whole sequence must be carried out in less than five minutes.

Surface decompression gives us the best control over the final phases of Willi's dive. He can breathe oxygen, move around, and be comfortable. We can control his depth much more accurately than in the heaving sea. If he has a medical problem, someone can join him in the chamber.

However, if we take longer than five minutes to get him from the water into the chamber, the chance of decompression sickness increases. It is a risk taken to avoid the larger hazard.

Willi's voice comes in low and quiet.

"Doc, I think I better warn you. I don't feel so hot. I'm not sure if it's the rolling sea or the dive. I feel sea-sick."

I stand up. In front of me is a difficult, and perhaps impossible, diagnosis. Is Willi sick because a microscopic bubble has lodged in his inner ear? Is it classic seasickness? Something else? Thoughts rise like larks scattering.

"All right, Willi. I understand." I try to mimic his calmness. "Think we'd better get you up and into the chamber on schedule. You're due to come up anyway in one minute."

"Roger, Doc. I'll hang in."

The problem passes soundlessly to the members of the deck team. The chamber is readied; in it are towels, dry clothes, and food. The main lock is at a pressure deeper than forty feet. The diver and his tender will climb into the entrance lock and turn an equalization valve. Air will flow from the main lock into the entrance lock. Both will equalize at forty feet. It is the fastest and safest way to recompress.

Willi rises from the water. He looks like some wild and wet barbarian warrior. His arms are flung wide and hold fast to the pipe frame. Water runs broken crystal rivers down his helmet and canvas suit.

The crane operator swings the stage gently to the deck. Big John sweeps in and takes a firm grip on Willi's helmet. He deftly spins wing-nuts and lifts the helmet free of Willi's head. Willi steps past John and pauses on a clear space on the deck. He bends over and throws up. He smiles at me weakly. He and I step into the chamber.

The hatch closes behind us. I reach for the valve. Gas roars in. The temperature soars. I feel like a fly trapped in a tobacco can fallen into a furnace. We rush to clear our ears.

With jet engine fierceness, we arrive at forty feet. Three minutes to spare.

Willi looks bone-pale. I help him from his gear.

"Christ, Doc, it sure happened quickly." Willi speaks slowly. "I felt dizzy and lightheaded. Do you suppose it was the air?"

Willi moves lethargically into the main chamber. He slips an oxygen mask off the wall and begins breathing. I watch him carefully. He seems unsteady, and shakes slightly.

My questions come without prompting. A memorized agenda from medical school.

"Do you feel any pain? Weakness? When did the dizziness start? Is it getting better? Worse?" An ongoing litany.

To each question, Willi nods or shakes his head, while holding firm to the oxygen mask.

I begin to examine him. A quick survey of his heart, lungs, and central nervous system. The cause of the problem keeps its face well hidden.

"It may have started when I switched back to air during decompression," Willi says. "The gas tasted funny."

A remote possibility flutters like bat wings in the night air. I reach out for it.

"Bill, will you have someone check the air compressor? See if it's picking up any carbon dioxide."

"Sure, Doc, I'll send Jim."

Willi improves. After a few minutes on oxygen his skin loses its old cheese colour.

"That feels much better," he says. "Sure is good to get out of those big swells. Myself, I think it was just seasickness. Strange kind though. Things seemed to spin around me."

A voice breaks in over the intercom. It is Kutzleb's unmistakable growl.

"You were right, super quack. The wind has changed. Might have blown some exhaust into the compressor intake hose. We'll have to be more careful."

Normally a ship at anchor swings its bow into the wind. However, *Rescue* was locked in a four-point moor. When the wind changed, the compressor exhaust gas was blown in a new direction. Some of it was probably picked up by the intake and might have caused symptoms.

In another two hours Willi will emerge from the chamber. He will feel fine, except for a trace of dizzyness.

But his problem will not entirely disappear. Despite drugs and pressure treatment, minor symptoms will continue to plague him for the next few days. He will suffer occasional lack of balance and unsteadiness of gait. It will take months to disappear completely.

I was never satisfied that we knew the real origin. Was it seasickness, carbon monoxide poisoning, or decompression vertigo? Or a combination of all three?

I glance from Willi, who is now relaxed and breathing easily. He and I both look out the porthole at the late sun glittering a path across the sea.

A thin, black line tightens beside the ship and then goes slack. It is holding something on the surface. It is the same line Willi took to the bottom. Its free end is attached to the body of a man. He has been on the seafloor for three weeks. He is no longer recognizable. The stench is incredible.

The days and nights slip by. Trade winds blow in softly from the south-east and occasionally carry sweet jungle scent over the ship. The sea continues to run in big swells beneath us.

The weather is almost ideal for diving. We ride high on a string of perfect days and make twenty-one dives. During this period we recover many critical sections of the aircraft including two engines, the tail assembly, and the flight recorder. Unfortunately, we lose a large part of the passenger section just as it is brought on board.

When a heavy object is lifted free of the water, it loses the lifting effect of buoyancy. A much greater strain is suddenly placed on the lifting wire. As we carefully ease the passenger section towards the railing, the thick line tears through the light metal construction. The airframe breaks away and returns to the sea. Only bubbles and a film of oil remain.

The two jet engines are a hazard to recover. They are secured on the sea floor by the diver and lifted to a point thirty feet below the ship. Another diver goes down and checks the original slings and fastenings. Adjustments are made and additional lift wires added.

The hard part comes when the cannon bulk of the engine is lifted free of the water. It inevitably starts to swing. Ten tons of metal in momentum. The ultimate wrecking ball. Dripping with oil and fuel.

I marvel at Kutzleb and his crew. They are the kings of wire rigging, and expertly lash and shackle the huge engine even as it sways. Extra lift wires and winches are added to the foray. With a unified voice, we curse and swear the swaying iron monster over the side. It crashes to the deck and is quickly lashed.

Each recovery operation brings its share of cuts and bruises. Many are too deep or severe to ignore. Occasionally, when a dive is not imminent, the pain is eased with a quiet tot of rum.

The whole of one rose morning is spent slowly moving the closed circuit TV through a prime wreckage area. Kutzleb and I stand together at the stern helping a diver check his gear. We look up and see a white-shirted, round shape running toward us. It is an accident engineer from the second deck, he trips and almost collapses down the companion-way.

"Bob, I've seen it! It's there! Right below us. We gotta get it!"

He pants like a dugong in heat. Kutzleb looks quizzical.

"Get what? What's down there?"

"The flight recorder! I saw it."

Kutzleb's wrench falls from his hand. In the echo of its clatter he sprints toward the bridge and the TV monitor. A few minutes later he emerges into the sunlight to speak to the diver beside me.

"Dale, suit up!" Kutzleb barks. "Go and get that yellow mutha."

Dale is built like a wrestler on protein supplements. He climbs into his tent-like canvas suit and makes a quick descent. We see him on the TV. He places a bright sphere into a bag and secures it to a leg of the TV cage. He decompresses and returns to the surface. After he is safely locked in the deck we begin to lift the cage and its precious cargo.

The flight recorder looks like an over-inflated yellow basketball. Under its cold steel casing is an impressive array of instruments. They monitor in-flight factors, such as time, speed, direction, G-force, and altitude. The water-tight integrity of the recorder preserves the critical history of the flight.

The imprisoned information is of special importance to this accident. Just fifty seconds before impact the pilot talked to the tower. He reported that everything was normal and asked if any other aircraft were nearby. "Negative," came the answer.

We are certain that the flight recorder holds at least some of the answers to the crash. Everyone leans over the rail to watch it break the surface. It is tied to the leg of the TV cage, loosely fastened inside a big canvas bag. As we watch, a big wave slides under the ship and swings the cage in against the hull. A hollow clang rings out. A strand of rope parts. Like an orange slipping out of a broken shopping bag, the recorder falls back into the sea. Lava boils over Kutzleb's lips.

"You goddamn weenies. Get down there and get that bloody thing. And for Christ's sake, tie it tight this time."

Big John is in the water almost before anyone can ask if he is ready. He is only briefly on the bottom. Within minutes he is back on the surface. It is one of our shortest dives.

"I don't think it will come loose this time, Bob baby." John smiles. "It may even try your Boy Scout patience." He secures the chamber door behind him.

The TV cage again breaks the surface. The flight recorder is tied onto the same leg. John's efforts look like work of a hysterical rope collector. A tight-fisted lash of hitches, overhands and square knots. Like nothing ever seen in the Boatsman's Manual.

Kutzleb takes ten minutes to cut the recorder loose. As he chops away, he mumbles something sounding vaguely like "sun and beach."

The flight recorder is shipped out on the next plane. In an isolated Washington laboratory, it will be studied by experts from the National Transportation Safety Board. We continue our work at the crash site.

Frequently our dives extend into the late evening. Some salvaged objects do not come on board until midnight. One reason for the delay is that we always wait until the diver is clear of the water before bringing up the lift line. This minimizes the risk of fouling beneath the ship.

Our objective is to recover more than just pieces of the aircraft. Twenty-one bodies are located and carefully returned to the surface. It is ugly work; both for the diver and the surface team. The bodies have been in the warm tropic water for three weeks. Decomposition is hideous.

Fortunately for us, the corpses are not taken aboard *Rescue*. Instead they are surfaced and towed to a nearby Venezualan Navy ship. This large, high-sided vessel stands about four hundred yards off our starboard beam. Her big square frame rides aggressive and loose on the swollen blue sea.

As soon as a body arrives at the surface it is my responsibility to ensure its safe transfer over to the Venezualan ship. Usually this is not a problem. One night it was.

It is our first body recovery, and it comes after a late afternoon dive. Other than some minor problems, the dive goes smoothly. The diver is tucked safely in the chamber. His pleasure at being successfully out of the sea radiates over the intercom. Under Kutzleb's direction the surface team gently eases the lift-line free of the bottom.

We have already seen the body on the closed circuit television screen. It resembles the swollen caricature of a middle aged man. It is strapped in a tilted seat on the sand.

During its upward journey the seat catches on a piece of wreckage. The seat belt tears open and the body floats free. Its interior gas expands and it arrives on the surface just off the stern quarter. In the gathering darkness I hear the hiss of gas escaping from torn skin.

"Lower the whale boat. Get Rafael and Billy. Better get going; the winds picking up."

It is the ship's captain. The crew moves smartly.

I grab some oilskins and scramble over the side into a small boat. It swings awkwardly and bangs into the bigger one. The bow line is thrown free. Rafael kicks the diesel engine into reverse. I feel its strong thunk-thunk under my feet. We swing out into the blackness.

Rafael stands strong and square in the stern. His dark muscled hand on the tiller gives him a fighter's confidence. Billy moves lithely about, readying lines in the bow. The boat picks up the rhythm of the sea.

We move slowly out of the twilight zone surrounding the ship. The free-floating body is on the opposite side, and we are forced to make a wide detour around the two mooring lines which loop down from the stern. When we arrive on the port side, the body has gone. Arms at the railing all point downwind.

The breeze spreads anxious fingers across the ocean. It is cool to the face. I look west. Somewhere, riding high on the waves, is a bloated human form. We head downwind.

The swells are huge beneath us. The whale boat roller coasts and lifts my stomach. An impartial moon hangs in the sky and silvers the sea. To the south I glimpse the soft haze hills of Tucacas.

It is difficult to see anything in the valleyed darkness. Swells rear black and soft beside us. We journey from one peak to another. I listen to the rhythm of my pulse.

The ship slips further and further behind. Soon it is only a glowing beacon. Rafael, Billy and I strain our eyes over the black hills. We see nothing. Rafael breaks the silence.

"Hey, man. Dis is goin' to take some time. Dat white thing is hiding. Our eyes is not much good."

He scans the rolling horizon. For the next twenty minutes we see and say nothing.

We crisscross back and forth across the waves. The only sound is the steady burp of diesel and the wind hiss of waves below us. Our eyes strain and receive only blankness.

"Why don't we go way downwind," I suggest. "Then we can make a long pass across the drift. Maybe our sense of smell can help us."

"Perhaps," says Rafael, skeptically. "Our eyes sure ain't doing any damn good."

We run west for ten minutes and then Rafael turns the rudder so that the wind cuts square across our beam. The whale boat doesn't like the new direction. It rolls like a pig in a trough.

115

Suddenly I am acutely aware of time and location, New Year's Eve. 1968. Just before midnight. My friends up north adrift in celebration. Warmth, light, and perhaps champagne. Bubbles and laughter on the tongue.

Then I smell it. Strapped in the arms of a soft breeze. Death. An aroma so hard in the nostrils that it is almost tactile. I clench my fists.

Rafael turns the whale boat sharply into the wind. The first wave splashes a white tip over the bow. Rafael steadies the boat and reaches into a pocket for his handkerchief. I do the same, knowing it will be inadequate.

It is an easy trail. We follow an aroma roadway over wavering hills. Within minutes, it is in sight.

It lies rank and bleached in a sea of occasional silver. It is a facedown arrangement of skin, heart, and brains becoming jelly. A stream of vile green fluid pours from an opening near the shoulder. The air jolts my throat and locks forever in my brain. Billy gags and throws up.

I reach down and pass a soft rope around the centre of the corpse. I gently cinch it. My mind moves in slow motion.

It is a long trip back to the ship. *Rescue*'s rails are deserted. Everyone has gone to bed. We slip quietly past the second hour of the New Year.

As we idle into the lee of the Venezualan ship, Rafael keeps looking back into the waters behind the body. All I can see are the glow wash of luminescent plankton. A following halo. No one speaks. Billy whistles softly to himself.

We transfer the body to the Venezualan crew. For a few minutes we stand off and watch. The men are incredibly inept at handling the body up the steep hull side. Its rag doll legs flop uselessly into the steel. The lift is excruciatingly long.

Rafael turns the whale boat back toward *Rescue*. The clean sea air washes my lungs. We near the ship. As I step off the whale boat I glance back and see what Rafael had been looking at. The triangle fin of a shark. Black on black in a silver sea.

Descent Nine
The Ice Lovers
1971

A huge block of ice jams its frozen bulk against my shoulders. Several smaller pieces float cold crystal near my neck. Streams of ice water trickle pain into my neoprene suit. I am afloat in a lethal sea of wet diamonds. I take a deep breath and shove hard. Suddenly I am free.

I am in the Gulf of St. Lawrence and surrounded for miles by ice. The nearest shore is a tiny cluster of sand braced islands—the Magdallens—twenty frozen sea miles to the south.

My diving suit holds me gently on the surface. It protects me against the deadly cold which could squeeze the life out of me in five minutes—if I lived that long. My camera hangs quietly in hand. Ahead is open water and the angry outline of tossed snow on a pressure ridge. Some three hundred feet below lies a sloping floor unseen by human eyes. Somewhere, hidden in watery folds between ice and floor, is my camera's quarry.

With my head down I can see nothing; nothing but sun-brilliant clarity. Then, softly at first, I hear a call. The muted melody sound of animals close to themselves and their newly born. A chorus of awareness—and perhaps of love. It is the cry of the Harp seal.

It is early March, 1971. This year, as they have for centuries, the Harp seals swam away from their northern feeding grounds and set unknown inner compasses for the pans of ice waiting for them in the Gulf. Their arduous journey has taken them past the mouths of glaciers and the wild sapphire loom of polar icebergs. They have seen the clustered white umbrellas of heaving seas and heard the terrifying fin-rush of killer whales. For thousands of cold night miles they responded to a single tissue flame; to seek and find the ice to birth their young.

First picture ever taken
of Harp seals in their
underwater environment.

The ice for miles around me holds hundreds of pups who slipped from warm wombs only a week ago. Their soft cries echo in the snow and the wind.

The sun tries to press into the water, but its weak spring warmth is lost in the upper layers. I move forward slowly, and set my eyes for surface level, trying not to disturb the sea.

Suddenly a dappled grey head; then another. Two skulls cascade separate sheens of water and turn quickly to look. A brief glance.

I am transfixed by their eyes. From each head two fathomless pools pulsate life with a vibrancy out of place in this chill ocean. I hear the soft sigh of twin exhalations and the throat-whistle of inhaled breath. Then they are gone.

I try to follow them. My head and chest drop forward and I give the full weight of my legs to the sky. My diving belt eases me down into a slide. A stolen breath from my regulator and I level off at twenty feet. I am cautious with each breath, for I know that any regulator can freeze up suddenly, forcing me to return to the surface.

I stop in an uncertain green room with no floor. Its quick-silver ceiling is rimmed with deadly lace ice. The home of the Harp seal.

In the clarity, ahead, is movement: a sinuating shape just beyond my vision. Its edges are blurred and gauzy grey. My fins drive me down and closer. Focus. A smooth form. Round, with head and shoulders tapering gently to a steer of tail. Two forward flippers scull and plane the water.

Effortlessly, the seal turns and fixes me with a wide-circled stare. It spins on some hidden axis, flank forms an ellipse, and is gone.

Then there are two. Gemini forms swimming everywhere— turning, twisting, above and below—seeming to move without energy. I am deeper now, surrounded by motion and speed. I am weightless in a darksome room that has no gravity. I am a neoprene Alice, peering through the looking glass. Ignoring cold, I turtle deeper into my wet suit. The twins continue their dance.

I marvel wordlessly, for I have seen nothing like it. It is a separate song and different from anything on land. It is unlike the tight leap or compressed twisting of captive aquarium animals. I watch a ballet of hot blood in uncaptured veins, and sea muscles running free.

Now they are above me. A pair of lissome shapes in ice-halo silhouette. They glide in chariots of rhythm pulled across a water sky. Then they are gone. They disappear into carbon depths below tomb-white ice.

I feel a sharp pressure around my waist. It is a tug from the yellow safety line which connects me to the surface. Its prodding confirms that I have used up its full downward length. I respond with three firm pulls—the signal to lift. On the ice overhead, two strong Newfound-landers begin the hand-over-hand that will transport me upward. I could swim, but I rest and take advantage of the elevator. I attempt to streamline my wrinkled bulk, but feel terribly ungainly. I have just become the first human to observe Harp seals in their own environment. I suddenly recognize that my own underwater movements are stiff and graceless. I am an awkward newcomer from the other side of the looking glass.

Within seconds I am surrounded by the surface world of block and brash ice. The ice is constantly moving and gives me concern. Any sudden shift could damage my equipment or cause injury. Sunlight and sea waves slap across my facemask. The taut line urges me through slush to a firm edge of snow. A strong hand lifts my elbow. I kneel at the edge of an ice pan.

Two fog streaming mouths open simultaneously. They pour questions through the cold.

"Did you see them? Did you hear them? How close did you get?"

A third face slowly fixes me with baleful stare reminiscent of the seal. His warm voice steams through his frosty beard.

"You look cold, me son. Don't kneel there like you're giving thanks for safe passage. Let's go get us some rum."

It is Farley Mowat. He's right. The water is already beginning to freeze-dry on my suit. The cold grinds into my pores. Gear is shouldered and we start the difficult walk across the ice for the ship. As we move carefully forward, I recall the day I first came on board.

"She ain't what you'd call a handsome vessel. Her sides are rusted and in places the ice has worn her thin. She's old and in need of paint. But she's got a good skipper and she handles well to the wheel."

A crewman from St. John's describes our ship the *North Gaspé*. He emphasizes his last statement with a volley of tobacco juice. It arches thinly over the side and down onto the ice.

"What's worse, a week after the hunt she'll start to stink like Holy Jesus."

I have been aboard the *North Gaspé* for two days. I am part of a CBC television crew trying to film Harp seals. My job is to get the first underwater pictures.

We have only three days left before the "hunt" begins. It is the annual slaughter-quest for the fur coats of thousands of new pups. Hopefully we'll be gone.

Farley and I reach the ancient ship jammed tight into yard-thick ice. Her engines are off and her generator reduced to a hum. She squats like a tired rusty monster on the roof of the Harp seals' home.

The *North Gaspé* is the shadow of death—for her straining bulkheads are the early spring dorm of the hunter. She is a sad vessel hinting of better times; summers at speed over windless seas, decks gleaming with cargo and laughter. She is a once-elegant ship, fated each March to transport ominous and tragic ides across frozen pastures.

I go below decks to change out of my wet-suit. After a shower I climb up towards the captain's cabin.

From the upper deck I see our solid berth in the centre of a huge pan of ice. Around us stretches a sub-arctic desert, except for the coal-grey shapes of seals hauled out on the ice. The small pups are almost invisible in their snow white coats; only their bleating and black noses give them away.

The elation of my dive slowly ebbs into sadness. In a few days the lower decks of the ship will disgorge a band of tough, hard-cursing Newfoundlanders. They will be bent on one thing—the killing of the pups. White fur will be exchanged for dollars and machismo.

It is a situation tumbling conflict through the brain. The "hunt" is not the simple problem many would make it. It is a complex web of abject poverty, hazards, and hardship, and the year's only chance at good money. There is the additional element of manhood. It is an injustifiable reason, perhaps, but one not easily dismissed. I have seen a parallel mystique in Alaskan Eskimos when they hunt the polar whale. Each man's search is centuries old, and will not die easily.

I am still cold. The welcome thought of rum adds spring to my step. I head towards the bow of the ship and the sealed hatch of the captain's cabin. I knock gently, step in, and close the door quietly behind me.

I have stepped into the warm den of Bacchus. Moist smoke and darkness compress visibility to zero. Asphyxia fingers my throat. The small room reeks of stale tobacco and 190-proof rum. Two of the inhabitants are unconscious; one open-mouthed on his back in the upper bunk, the other with chin on his chest and about to fall out of the same bed. Fortunately there is no roll to the ship, or he would end up in a round, white sink below.

It is a small room with two bunks, a desk, and two chairs. The other major features are a port-hole, vainly trying to drive sunlight into the blackness, and the tiny sink in its depressing role as fireman's net. A gallon jug of hard black rum sits squarely in the centre of the floor.

My slowly-adjusting eyes make out four dim figures in addition to the two under anaesthesia. All four smile wickedly, with grins framed by the blackest of beards. High noon, dead centre in the Gulf, and a virgin has stumbled into the cavern of black infinities. I shuffle toward the door mumbling, "Sorry, I must be mistaken. . . ."

Laughter rolls out of the nether regions of the lower bunk. "Ha! So you're the idiot who goes underwater with the swiles. It's amazin' anyone could be that crazy and live to tell about it."

The captain's voice booms, "Now sit down me bye, and join us for a short drink. We want to hear about the swiles."

I stifle the desire to cough, and find a seat on the lower bunk. Two of the other occupants are Mowat and a young gentleman of eighty. The breath of rum is so heavy that a drink is quite unnecessary.

Farley leans over and whispers, "What did you expect? The Christian Science reading room?"

A silent initiation ceremony begins. It is conducted by the captain who sits in a chair opposite the berth. He reaches down for the gallon rum jug. I shudder. I have never seen a potable liquid so dark. It looks like bottled smoke.

"This is what we keep on board in case one of the boys falls through the ice. If he survives the water and the trip back to the ship, this old fire brings him round soon enough."

"Certain, you'll share a wee drop with us before lunch."

I nod slowly.

"Well, since you're such a young lad, and fresh from the sea, I think it best if we cut it a little."

I breathe relief. The thought of downing the black acid heat trembles my liver.

The captain reaches for a plastic tumbler, shoulders the jug, and splashes the smokey liquid almost to the brim. Again I shudder.

"Ah yes, let's cut the bite down with a little of this."

He reaches over the table and curls a big knuckled fist around a bottle of Bacardi. The light green liquid splashes and bubbles on top of the Demerara. The tumbler is now full to the brim. Pure rum. I am almost prostrate with anxiety. A harbinger of positions to come.

At first I think I can avoid drinking the mixture. I discover I can hold the tumbler below my nose and simply inhale. When my nose begins to get numb, I take a small sip. After the initial violence to my throat, the liquid doesn't taste bad at all. Soon my whole body begins to get numb. Slowly I slip back so that the bed springs of the upper bunk are in focus. For a while.

I never get to talk about the seals. There are two reasons. For the next two hours I listen to non-stop honest yarns of past adventures along the coasts of Newfoundland. There are tales of ships and fogs, rocks and waves, and the incredible courage of men. I have good reason to keep my lips closed. I am listening to history. And its makers.

The second reason for my silence is related to the accident spreading across my face. A unique form of lock-jaw invades my throat and my tongue. My nerves and muscles are partially anaesthetized.

I don't recall seeing the bottom of the rum glass. I do remember standing with some difficulty and a hand of gentle assistance to the door. The help is rendered by a gentleman twice my age. I also remember the blinding crash of sunlight outside the cabin and the surgery of the sea air.

There is no more diving that day. All the ice-leads within miles of the ship quietly close and lock up the sea. Fortunately. By coincidence it is the same afternoon that I have an overwhelming desire to lie down and check my eye-lids for light-leaks. I can't find any.

The next day Farley and I walk alone out over the endless ice. It is an unearthly platform that seems stable and almost permanent. In reality it is a temporary collection of plates and hummocks adrift over deep black water. Cracks split open with the suddenness of a rifle shot. To walk these fields is to risk separation from the ship and unwelcome entrance into forbidden crystal caves. We move slowly.

A light snow falls intermittently. Behind its diffusion drifts a watery sun. It is just visible.

We walk in a huge pearl hall carpeted with hard sugar. It is not the best time for such travel. Snow can mask weak and pitted ice. We step carefully and keep the outline of the ship firmly in view.

Fortunately we do not have far to go. Just ahead lie half a dozen seals clustered next to three round holes in the ice. The holes are four feet across, and edged with round, smooth shoulders. In the event of danger, the timid animals disappear down these open gates.

Seals lying on ice are all bulk and belly. They are torpid pears who have cast away the torpedo elegance of the underwater world. They move by digging in their foreclaws. Then, in humping caterpillar fashion, they drag up their sprawling hind-quarters. This ungainly and comical movement is surprisingly fast. It is always accompanied by the rasp of sharp claws and the slither of ice-coated fur.

Farley and I approach cautiously. We drop to a semi-crouch, and move ahead in single file. Swiftly, without caucus, the three seals nearest holes slosh into the black matte. Farley and I lie down quietly behind the grotesque ledge of a shallow hummock.

Our disappearance brings the three females slithering back onto the ice. They slide into wet closeness with their pups, and present milk-heavy breasts. Unbroken wails cease as the pups nuzzle and draw, their mouths moving in mild frenzy.

Farley looks along the jagged length of the hummock.

"Amazing thing, this white universe. Who would believe such acid cold could contain so much life? The green liquid below this floe has swarms of microscopic animals. Countless fish and, for a while each spring, this huge population of seals. God knows what else is down there. Perhaps killer whales and great white shark." Farley looks into the dark holes.

"It's easy to understand why life began in the sea. There's vitality in every cubic inch."

"What is so sad is that most people don't know what's going on out here. The sea is what sets this land and this planet apart. Most people don't give a damn. We dump poisons in it and slaughter its inhabitants. We don't know it, but we are killing our biological history."

A new seal appears out of a hole. She is alone and carries a long slash wound across her chest. Blood drips from torn skin. The new-comer skitters and sniffs across the ice to a mewing pup.

A dissonant roar is unleashed. The pup's mother arches a massive chest and hump-scrapes toward the intruder. Claws dig like talons into the ice. Another roar crashes from lungs. The wounded seal retreats. Her search for her lost pup continues elsewhere.

We stay with the seals for an hour and watch them suckle and nurture their young. We share a window into ancient harmonies; loving animals in precarious balance with nature.

On the way back to the ship we stop to look down a fresh seal hole. We stretch out full length on the ice and stare into the unpierced, unreferenced, zones below. A chorus of faint silver notes reaches up from the water below my eyes—but no, it is only a hoped-for, and imagined, sound.

The sun breaks the cloud tatter and sends swift slanting rays into the pool. We gasp at the fleet grey shadows sweeping across the bottom of the black grotto. At least a dozen or more. Occasionally a shadow loses its vagueness and we see the harp shape outline on its back. We hold our breath at the celebration of life.

Farley looks up and gazes across the ice of the frozen sea.

"What incredible harmony. Harsh and yet serene. Out here you can see how wrong we've often been. We spend so much time grasping for material things. The result is that we've lost, and no longer feel we need, the basic life rhythms. A terrible loss for man."

We walk back to the ship behind separate walls of unspoken ques-tions. We board by climbing up the hull on a rope and wood ladder. It is a difficult ascent. I am alone with iron thoughts.

The day has turned to evening and the ship echoes no sound. It is tomb quiet. Tomorrow the hunt begins.

But what of the ice lovers?

Descend with me for a moment into their black green universe. Re-focus your eyes and mind on an animal in his unknown home. Fly and follow him into polar water so clear it vanishes. Pursue a path into honeycomb caves and labyrinthine channels. Feel the python grasp of cold against your frail neoprene armour.

Look now through double twilight along moon white ledges. See sunlight cascading down broken staircases of ice. Listen for a moment to the stirring call of a new friend. Notes sung by the lithe chanticleer you are beginning to love.

Swim into a cave, with plankton clinging like yellow gossamer to its ceiling. Enter with care, for the walls slope with the strain of sudden construction. The cave may collapse with the next heavy wind.

Rise through flowing jellyfish, stretching tentacles away from sunlight. Watch for flicker fin in the sea ahead. Be wary. Relentless teeth of predators are below. Hiding somewhere beyond vision is the torn sneer of a white shark. The freshly-killed seal behind his jaws will not quiet his belly for long.

Ignore gravity. Remember the whales who recently swam through these weedless halls. Huge placid animals, who knew the Ice Age; who have learned from man the shriek of steel into flesh.

Look at the seal you are losing beneath these canyons of ice. Wonder at those who would kill such sweetness.

Ocean Systems diving team
on board a Coastguard ship:
diver is preparing to enter the water.
Author (right) walks aft to inspect
life-support station.

Descent Ten
May Day In Michigan
1971

The night air screams. The pulsing sound of steel rutting into wind. Over my head invisible helicopter blades scythe the falling snow and slap out a deadly morse code. The shouting tips travel at hundreds of miles an hour. I hunch down and continue toward the open black doorway.

The engines drum hard against my ears. The taramac trembles under foot. Its wet ice is slippery as oil on butter.

To my left is the killing blur of the tail rotor. Its blades spin a deathly silhouette.

A pale face peers at me from the blackness of the doorway. A hand reaches out. I grasp it and step up. The door slams behind me. The noise of the maëlstrom dims.

"Sit over there." A nameless voice shouts into my ear.

"Strap yourself in good and tight. It's blowing like hell upstairs."

I fumble my way into a small seat beside a window. The darkness and bulk of my parka hide the seat belt. I find the buckle next to the floor. Its solid metal lock gives little comfort.

My mind throbs in time with the blades. Only a few moments ago everything was normal. I had spent the past few days at the Ocean Systems Research Laboratory in Tarrytown, north of New York. This morning I flew to Detroit and then into northern Michigan to join the Ocean Systems search and salvage team. They were to make their first dive but I was late. Because things had gone well for them, the first dive had been moved ahead to today.

The dive was routine until it was time for the diver to return to the surface. In the dark void at two hundred and fifty feet he became entangled. Suddenly he was a weightless prisoner suspended in blackness. Trapped in wreckage.

The helicopter tilts up and forward. Its engine groans. Snowflakes blur. Blades scream purchase. The helicopter lurches up, as if hitting an invisible chuckhole. I take a deep breath of the cabin's cold mist.

The telephone call came about an hour after I stepped off the airplane.

"Dr. MacInnis? We've just had an urgent call from the salvage team out on the lake. One of the divers is fouled in wreckage. His name is Maltman. They need you right away."

A cold ember burned deep inside me. An ice-nugget of fear.

Tangled in the wreckage. Bill Maltman. An old friend of mine. An experienced diver. I picture him cold and weightless in the water column. Surrounded by blackness. Somewhere taut below him lies his thin breathing hose twisted through sections of broken metal. Overhead, 250 feet of freezing water. His decompression obligation mounts. A counterfeit move can sever his breathing hose.

I picture Bill Maltman forcing calmness down his own unsteady throat.

The ember moves. I shiver. I don't like the grim images of death fluttering around my friend. Evil wings loose in the water. Hazards contained and suddenly loosed.

The telephone was adamant.

"A helicopter is standing by. As soon as you can get to the airport we'll fly you out."

Stars dance under the snow. I see the lights of the town. Charlevoix, Michigan. Summer vacation retreat of the upper peninsula. Refuge of the rich from Detroit and Chicago. Not far away is a United States Air Force Strategic Air Command base. In early January, a four-engined B-52 was on a low-level practice run over the lake. A thousand feet up, and moving at six hundred knots. Without warning the huge plane plunged into the icy waters.

The cause of the crash is a mystery. Fortunately no hydrogen bombs were on board. All hands were lost.

The wreckage was strewn in a long, broken pattern across the lake floor. Crazy pieces of engines, wings, and fusilage. Everything is guarded by an angry barricade of torn and broken aluminium. Razor-sharp, in the wet perpetual night.

I marvel at the swift change of events. One minute, there is an extraordinary strong and complex airborne system. The next minute, there is a spray of exploded metal in cavernous waters. The pain is fierce and unyielding.

The helicopter pitches and yaws in the harsh night air. Danger dials tremble. I shrug deeper into the chill warmth of my parka. The helicopter gyrates wildly. Outside is a cold, dark wind. I fly lonely within myself, and in a deep chamber of suspense.

The town gives way to forest. Dark swatches of evergreen shag crazily across white ground. Snowshoe country. Hiding place of deer and timber wolf.

The helicopter flies low, seeking landmarks. The snowfall is lighter now, but the turbulence is not. Heavy fists of air lash out at the helicopter. I run my fingers lightly across a line of cold rivets next to the window. Millimeters separating flesh and sky. Delicate and temporary balance.

Forest turns into seawrack. Long lines of white breakers roll up and over the fast ice. The lake is angry with wind. Icy spray showers across the uneven shoreline. The background waters are black and swollen. We head out over them. The engines continue to scream.

Lake Michigan. Part of a huge inland sea. The Great Lakes. Enormous glacier-carved basins. Ninety-five thousand square miles of restless water and ice. Tonight's terrain.

Bill Maltman's life hangs on several threads loosely held together. One of them is a special diving suit. Hot water is heated in a small boiler on the ship high above him. It is pumped down through a hose, which joins his suit at the waist. Here it passes through control valves and a series of tubes and orifices. Then it washes out over his skin. He is suspended in a bath of hot water. Epidermal delight in the darkness.

The suit was built by a genius in California. It permits divers to work at unheard of depths and durations. Its thermal comfort is a splendid conspiracy against cold water.

However, if Maltman's suit stops working there is sudden and serious danger. The lake water would drive arctic cold through the close-fitting garment. It is a cold so severe that it would completely disengage Maltman's central nervous system.

My mind flicks through the memory of one diver who inadvertently shut off his inlet valve in the same cold water. Within seconds his voice was an electronic smear. I heard the broken sound of a man grabbing for breath. In less than a minute he was on the surface. He could hardly move. His flesh felt like seasoned wood.

I see the ship. It glows like a faint miner's lamp. It drops away. The waves lift and surround it. It is a soft glow blinking and beckoning in black.

Lights go on outside the helicopter. White on white. A thin blizzard of magnesium flakes. Ersatz Brownian movement.

The engine changes tone. Lower. Condescending. We slow our forward rush and pitch sharply down.

I am concerned about my exit. We certainly are not going to land on any quivering ship's deck. I don't like the other options. My sweating palms agree.

A nameless figure moves down from the flight deck and begins to work at something resembling a human stretcher. My invalid eyes follow his every move.

He turns and reaches for my parka-hidden ear.

"Look. It's easy. When I give the signal, you lie down on the rack. They'll recover you on deck. Just hang tight."

My fingers curl white and damp around the seat posts. The figure strains and the door groans on its track to reveal a black sepulchre opening.

I breathe deeply. O-Oommmm goes the engine. On goes my life-jacket. I hesitate myself into the stretcher. Doctor or patient? Its chicken-wire outline has no substance. I cringe. Firm canvas straps are secured across my legs and waist. Reluctance overcome.

Suddenly I am airborne. Hanging free in the wind-scream. Hopeless mummy staring into the night. Heart on its own wild gallop.

The wire cable inches out. Slowly. Painfully. The helicopter recedes. Reality wobbles.

I am midway between the chopper and the ship. A wind-blown seed between two man-made things. I start to pendulum. At first, a waxen swing. Now below the yellow mast of the ship. It assumes the motion of a tree falling over. Repeatedly.

Hands reach out but I glide gracefully out of reach.

"Peter Pan, at our service," a voice calls out.

My eyes squeeze shut as the stretcher swings in tight to the mast.

"Quit screwing around, MacInnis! Get your ass down here so we can go to work."

It is the metallic bark of Bob Kutzleb. Carried softly on the wings of the snow. The voice in charge of the operation. A born leader.

Another swinging pass. Lower and faster. Hands reaching up again. Like adoring groupies at a rock concert. I hope for contact. It comes. The cable slacks and I fall on the slush-covered deck. Pan loses his wings. A flopping, stranded fish. Voices converge. Faces surround.

"Jesus, some guys go to any extreme for a dramatic entrance."

"It's the Flying Nun from hog heaven."

"Pray, dear fellows. The saviour is delivered from the sky."

I carefully adjust my heavy sherpa hat. I am among thieves and scoundrels. Time for dignity and decorum. The actor emerges. I switch on my past-the-graveyard whistle. It limps. I walk away on new-found legs. They tremble.

130

Memo to central casting: About the Flying Nun. Next time, cast someone else!

Humour becomes ephemeral. It's a delicate bubble in a storm of seriousness. Frowns gather like angry cumulus. Kutzleb speaks quietly.

"He's still below. At ninety feet. It's been a hell of a mess—but I think things are now under control.

"We're not exactly sure what happened. Seems like Bill got fouled almost as soon as he started working. He spent about twenty minutes on the bottom trying to untangle himself. When it was time to decompress, he was able to climb up the main guide line to 160 feet. Then he got even more fouled. Things got worse when his depth indicator froze up. We decided to lift everything a measured distance and continue decompression—but somehow Bill drifted free to the surface. We almost had heart failure because of his decompression requirements. We zipped him back to ninety feet."

Kutzleb's voice clicks out facts. It has the confident ring of a man who has overcome an almost overwhelming challenge.

"So far so good. I want you to talk to him, Joe. Then let us know how we should decompress him."

I breathe relief. Events are still in Maltman's favour.

Kutzleb continues.

"He's been in the water for ninety minutes. We've got to get him the hell out."

I review the history of the dive searching for concrete details and facts. The dive log makes me shudder. After thirty-six minutes, Maltman suddenly left 160 feet. An uncontrolled ascent, almost to the surface. He recovered quickly and pulled himself back down. The unexpected trip puts an unknown warp in his bottom time. Gas embolism? I picture bubbles expanding against Maltman's delicate lung tissue.

I move in close to the intercom. Its silver toggle switch is cold to my fingers. "Bill? How are you? It's Joe."

"Doc? Hey, how the hell are you? The guys told me that the big metal stork was delivering you."

Maltman's light and calm voice masks his fear.

"Bill, I want you to give me a complete picture of how you feel. From top to bottom."

Bill, old hand, has been through this before. He knows what I am looking for. He runs down the list and hides nothing. But his final statement is awash with fatigue.

"In summary, Joe, I'm really tired all over. My arms are especially sore. Feels like twin charley-horses."

Maltman doesn't mention fear. He doesn't need to. We both know it's there. Probably deeper in my bowels than his.

"Okay, Bill. Hang tight. You're in really good hands. I'm going to take a few minutes and come up with a new decompression schedule."

I scud across the deck and slip into a small quiet room. Its hot bulb blinds me. The sudden heat is seductive.

I lay out all the figures. Compression time. Bottom time. Unexpected ascent. Return to depth. Decompression so far.

My pen gathers momentum. Columns of numbers and times. Optional decompression schedules appear from my briefcase. The alternatives narrow.

I weigh and decide. Time for the right answer. Maltman's life hangs in the balance.

I leave the room and move forward along narrow corridors to the bridge. Within seconds the call is going out to our Tarrytown laboratories. Shortly a voice answers. Its firm Austrian accent is comforting.

"Heinz, I have a decompression schedule I want to check with you. Can you give me a minute?"

I explained the situation. Heinz' mind whirrs with computer speed. His scope is prodigious.

"I'll have to have more time to check. Sounds good so far. Go ahead and start. I'll get back to you."

I feel much better. The decompression battle has been joined by the best brain in the business.

I race back down to the dive control area. The snow blows torn pockets of white against the deck lights. The wind bites hard against my face.

We lift Maltman up ten more feet. His body agrees. No new signs or symptoms. The suit continues to protect him with its vital warmth—its heater and hose system a quiet miracle. In any other apparatus he might die of exposure. I glance over to the middle of the deck. The hot water pump and boiler continue their indifferent mechanical songs.

Time passes. Maltman continues to rise slowly toward the surface. We all stand and wait, watching the seas.

I listen to the intercom and occasionally say a few words to Bill. His response is always calm and measured.

The time arrives for surface decompression. Maltman is to ascend from forty feet to the surface and be recompressed in a chamber below deck. It will require split-second timing, for it must be completed within five minutes.

The cadence begins. The diver is near the surface. Up to the ladder. Up to the deck. The arms of the line of tenders swing. Swift but gentle the motions. Helmet off. Hoses uncoupled. All hands for the diver.

Faces crowd in to help Bill. He looks pale and tired. His eyes glance at me through the blowing snow. He moves across the deck.

I join him. We talk. He steers the conversation away from himself.

"Damn strong wind tonight. Just like a storm on the Gulf. Except for the cold. Feels like the bloody Arctic."

His pace and movement are steady and poised. I follow his big form down the stairs. He steps into the open belly of the big iron chamber. I follow. The iron hatch slams shut. Pressure roars. Within seconds we are at sixty feet.

Maltman's miracle continues. He remains free of symptoms. I look at his arms. They are bruised and trembling. I picture them working furiously in the deep water blackness to get free.

Somewhere beneath us, the ship's engines hum. Our mooring has been cast aside and we are heading into port. The two of us roll hard against the curved white wall of steel. A big wave lifts the ship and then another.

It will be midnight before Bill Maltman will leave the chamber. Two hours later he will return with pain running through his elbow. Delayed decompression sickness. We will treat him again. Four and a half hours later he will emerge. The sun will be edging over the Michigan treeline.

He now leans against the wall and grins. No matter what the circumstances, this is a man always astride the edge of humour.

"Doc, sometimes I don't understand you at all. You could have a nice, quiet practice in medicine. Sit back and take in the big bucks. So how come you hustle out here on this bad bitchin' night. Just to chase after some sloppy diver boy from Morgan City?"

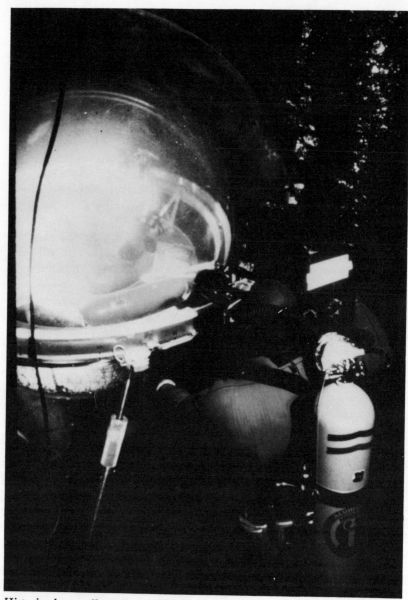

Historic phone-call: Author and Douglas Elsey
inside *Sub-Igloo* just prior to calling the
Prime Minister. *Sub-Igloo* is filled with air;
Rick Mason stands by with motion-picture camera.

Descent Eleven
Diving Beneath Arctic Ice
1972

A filigree of ice begins to form inside my face mask. I lean forward and let in a trickle of seawater to swish the frost away and clean my lens. The frigid brine burns painfully across my skin.

Ahead, faint in the seemingly celestial light fringing the work area, a transparent plastic dome rotates slowly on the tension of barely visible control lines. Then it begins a curved descent toward the sea floor—an adagio movement that hints of the touchdown of some ghostly spaceship.

A second diver silhouettes in the blue-black water and gives the dome a gentle shove. It swings over to a large, bright aluminum ring supported by 16 silvery struts attached to ballast trays. Beneath the ring another plastic hemisphere, partner to the dome, is already in place. I am looking at *Sub-Igloo*, about to become our new underwater workshop 35 feet down in Canada's ice-covered Arctic waters. *Sub-Igloo*, one of the most inventive technological advances in two decades of under-ice diving, will be the first manned dive station in the northern polar sea.

In the cone of golden light beneath the three-foot-thick ice overhead, the diver struggles to join the two hemispheres in a watertight fit. Beyond the electric radiance of *Sub-Igloo* lie the endless midnight hallways of Resolute Bay.

For a moment I am caught in an acute awareness of our location. Resolute is almost 600 miles north of the Arctic Circle. If I could extend my gaze south for a few miles, I would see into Barrow Strait—the northern corridor of the fabled, ice-choked Northwest Passage.

We are only 125 miles from the north magnetic pole. The rest of the world's compasses point toward it, and yet our own rotate endlessly and uselessly, seeking a pole that is confusingly near. We are so far north that a week ago I stood on the ice outside our tent and saw the northern lights in the southern sky!

We have come here at the calling of an unknown dominion—the northernmost continental shelf. It is so immense that in Canada alone it covers almost a million square miles, but human eyes have seen less than a few city blocks of this huge maritime estate.

There is much to be learned in these wet polar corridors—about arctic marine life, underwater ice structures, the composition of the bottom, the extent of pollution, the possible existence of new resources. But men must first learn more about diving, and surviving, in what is certainly one of earth's most inhospitable environments.

The ice world above us is no less daunting: no sun during the day, temperatures to 45 degrees below zero Fahrenheit, punishing winds. While we were putting up the dive tent, the wind suddenly gusted to 35 miles an hour; that gave us a wind-chill factor of some 80 degrees below zero.

One of our team, Dick Birch, is from the Bahamas. "If that wind had been blowing south, I would have used the tent to set sail back to the islands," he says. "Working on this ice and wearing all these clothes is back-breaking. I never would've believed any place in the world could be so unearthly cold and black." Dick stays, of course, like the rest of us, convinced that what we are doing is worth staying for.

The chief purpose of our study is to come to grips with the problems of scientific diving operations in the Arctic and to test both equipment and human performance. When we finish, we will have made more than 200 dives and studied problems of suits, helmets, breathing gear, a diver-propulsion vehicle, and even watches. I have a particular interest in learning just how much a skilled diver can attempt in these waters, where the temperature sinks to a subfreezing 28.5 degrees Fahrenheit, and accidental exposure can bring death in as little as five minutes.

As I am about to join the diver who is assembling *Sub-Igloo*, a clear Arkansas drawl reaches my ears without benefit of headphones. It is Chuck Cantrell, our topside supervisor, talking to me via a new underwater speaker that radiates sound 300 feet.

"Once the dome is in place over the hemisphere, give us two tugs on the line, and we will give you some slack," Chuck tells me.

I duck my head into the air bubble of a nearby Sea-Shell, one of our four plastic communication-refuge stations. I pull my mouthpiece forward and speak quietly into a microphone floating on the icy water. "Okay, Doug is just centring the two hemispheres at the equator. I'll let you know." Doug Elsey is an ocean engineer, and my buddy on this dive. He is one of 15 diving scientists I have brought with me into this severely beautiful but awesomely hostile ocean.

Now a new voice comes through the darkness. It is Birger Andersen, who directs the expedition's human-performance program. "When you two are free, would you return to the dive hole for a body-temperature check and status report?"

"Roger," I reply, and am reminded of the electronic pill I swallowed yesterday. Its function is to measure my "core" temperature from deep inside my body and send out a continuous radio signal. But the batteries had been weakened by the cold before I swallowed it, so I must return to the surface often to have my temperature read. Any drop in core temperature is cause for concern.

For this dive I also wear a device that senses my heart rate and pulses it through the water to a surface receiving set. As expected, the searing cold underwater upsets this delicate wireless equipment, and sometimes prevents the physiological signals from reaching the surface. But we generally get enough data recorded to confirm that our bodies continue to work normally behind the thin cushion of air held by our suits. What is not measured is the suppressed anxiety we all feel about the hidden ruthlessness of these cold black waters.

It is time to give Doug a hand. I duck below the aluminum rim of the *Sea-Shell* and push off. Because I want to be slightly less buoyant, I release a stream of air from my suit. For good reason, I wear no swim fins and, as I "moon-walk" across the sea floor toward *Sub-Igloo*, my feet settle softly into the amber sediment that boils up into small thunderheads. We have discovered that wearing swim fins near the bottom creates large dark thunderheads that reduce visibility to zero. Thus I dance finlessly forward like the helmeted diver of time past. Later in the expedition, former Astronaut Scott Carpenter will join us for a few days, and I will watch him take this same "weightless" seabed walk.

As I near the glowing *Sub-Igloo*, an exotic jellyfish with garish "neon" skirts drifts into the light. This undulating animal is a living symbol of these ocean pastures. When I brought my first expedition to Resolute Bay in 1970, I was amazed at the vast numbers of sea animals and plants on the floor beneath the ice; like most people, I had expected to see an underwater desert.

Our biologist, Dr. Alan Emery, was also surprised on his first dive. "I found plants and animals more abundant than I ever expected, though compared to the tropics, these waters have much less variety. Without actually diving into the cold depths, we would never have realized how plentiful Arctic marine life really is, and yet how painfully slowly it grows, moves, and reproduces."

I reach *Sub-Igloo* and begin to work beside Doug, fastening the bolts that clamp the domes together. We have been underwater for almost an hour, and my hands ache from cold. I slow my breathing and try to work smoothly with my wrench and bolts. It is not easy with my fingers in half-inch-thick soft rubber gloves. But it would be impossible to work bare-handed in these waters. In seconds hands would become stiff and feel as if they had been slashed by jagged iron.

In spite of the cold, we press on. The triumphant moment when we fill *Sub-Igloo* with air is tantalizingly near. But it will also be a dangerous moment. *Sub-Igloo* will become a giant bubble trying to reach the surface with an upward force of eight tons. We hope the eight tons of ballast in the trays anchoring the struts will keep the whole structure from roaring to the surface.

We must also be careful not to break our bubble. As Doug, who assisted in *Sub-Igloo*'s design, once said, "There will be tremendous potential energy held captive by that fragile-looking globe. If we drop a heavy tool or weight belt on it, it might shatter."

If *Sub-Igloo* works, it will be the world's first diver-assembled manned station in the Arctic. It requires no heavy lifting equipment to handle, and our divers can readily take it completely apart underwater and move it to a new position.

Sub-Igloo is like an explorer's tent. It provides the same kind of base—for storing our equipment, communicating with each other, and providing an easy refuge for a diver in trouble. More important, it is an extraordinary window on the underwater world. We can sit inside —comfortable and relatively warm, free of our breathing apparatus—and study the ocean floor, the water envelope around us, and the ice overhead.

Suddenly Doug and I look at each other. It is one of those unspoken and unpredictable communications that frequently occur between divers. It is our sixth-sense way of overcoming our inability to speak easily to each other underwater. We nod agreement. It is time to surface. We are both cold.

Doug motions toward the ice above us with his thumb, and shared laughter echoes faintly behind our face masks. We both know it is "upside-down" time. Time to stand on our heads and walk on the underside of the ice to the dive hole.

Doug and I lean away from *Sub-Igloo*, fall gently backward, and depress a round valve on the front of our inflatable suits. I feel a soft hiss of air on my chest and the beginning of an effortless buoyancy.

From below, the massive ice barrier, flat and almost featureless, resembles a faded pearl ceiling. Huge clusters of our spent breathing air compressed against it reflect the lights like thin pools of silver-blue mercury.

I realize I am going too fast. I empty air from my suit. Too late. Legs and arms outstretched, I hit the ice with a balloon-like bump and set off a soft explosion of displaced crystals. I let out a little air and kneel on the flat underbelly of the ice.

I am about to stand erect, head downward, defying gravity with the aid of buoyancy. Slowly I orient myself. The ice becomes a floor, and the seabed becomes my new ceiling from which *Sub-Igloo* seems to hang like a chandelier.

The smooth ice floor is broken only by the glow of warm light from three 100-watt lamps suspended near the dive hole. But Doug and I are not alone here. Living animals are embedded in the loose, complex crystals on the underside of the ice. My light picks out two copepods clinging together. Their light-brown bodies are motionless, and their inarticulate clinging underlines the crushing cold of this great lonely sea.

Suddenly my ears awaken to an inexplicable roaring. I wonder if this is a prelude to dizziness or nausea. Did I come up too fast?

Doug's eyes are wide and concerned and I suspect he too has the same problem. The sound changes tone and begins to drop. Then I realize it's not in my head, but on the ice. We have stopped just below our tracked vehicle, parked only three feet away, with its engine running. Because water is an excellent conductor of sound, noises can be heard with frightening clarity, even over long distances.

From my kneeling position, I attempt to stand with my feet on the ice above me. It is not easy. I feel awkward. The legs of my suit fill with inrushing air. My 50-pound weight belt shifts slightly toward my head. Any quick action will tip me over backward.

All at once I am standing—upside-down. The ice stretches out ahead and fades into a blurred horizon. To preserve my equilibrium, I move forward with awkward stiffness.

As Doug and I begin a wooden shuffle toward the dive hole, there is a flash of light. It is photographer Bill Curtsinger filming animals just below the ice. Slowly he rolls into a ball, places his fins squarely against the ice, and straightens, so that he too is hanging upside down. He begins to focus his camera on our bizarre walk.

We are something to see in our bright red suits, hanging in the pale light, our legs swollen to elephant-size with air. Our exhaust bubbles stream incredibly toward our feet to splash on the ice and flatten into thin disks.

In a burst of exhilaration, Bill begins to jump away from the ice. We all laugh, discover a new game, and begin to bounce like three red bears with springs hidden in our feet.

Then I see the square outline of the dive hole and the warm yellow ceiling of the tent. Three faces stare through at me. The inviting opening is below me. I dive into it, headfirst.

Suddenly I am re-oriented. A blaze of light, and voices greet me. It is the topside team welcoming me back to the realities of the surface. Phil Nuytten reaches down to help me. He is the surface supervisor for this dive. With water streaming off my suit, I slide up onto the glistening wet floor of the dive tent. I push off my mask and a wave of warm air caresses my face. Someone strips off my gloves and pours warm water over my numb fingers. A cup of hot chocolate is offered, and I gulp down its heat.

The surface team moves smoothly and quietly, assisting me in every way. A fresh tank of air is slid into my harness and someone adjusts my face mask, readying me for the next dive. We try to have at least one man on the surface responsible for each diver in the water. To give everyone as much experience as possible, we change roles often. Today's diver becomes tomorrow's helper or supervisor, and vice versa.

Doug and I sit side by side at the edge of the ice hole, gaining back some of our lost heat. We take advantage of the break to check over some of the ongoing work with our teammates. Tim Turnbull, a doctoral candidate in marine biology, and one of the two men helping me with this dive, wants to go under as soon as we have put *Sub-Igloo* together. He is making a photo survey and specimen collection, concentrating on the symbiotic relationships of small ocean animals. While he is down, he plans to do some checks on our closed-circuit breathing system, a very advanced kind of underwater equipment we are testing.

Roger Smith, the expedition geologist, is laying plans to take more core samples of the soft sediment near a kelp bed. Ches Beachell, the mechanical genius who has set up all our lighting and communication equipment, is wiring phone connections to *Sub-Igloo*.

Doug and I, of course, are planning the final assembly of *Sub-Igloo*. "We'll just tighten a few more bolts, and the hemispheres will be secure," I say. "Then we'll fill her with air and slip inside."

Doug looks worried. "We will *if* the bonding at the equator, between the plastic and aluminum, holds. It's never been tried before at this low a temperature."

Our conversation is interrupted by a geyser of bubbles and water in the dive hole. A diver's head and shoulders shoot up and out of the foam. It is Rick Mason, who has been filming the highlights of the expedition, mostly underwater, for the National Film Board of

Canada. Quickly he flips back his mask. His breathing is fast and laboured.

After a while he explains. His regulator froze in the closed position and he made a fast emergency ascent, venting the expanding air in his suit. That's what caused the geyser. "Sorry for all the splash," Rick says. "I'll get another regulator and get back to work."

His calmness is impressive. Like all of us he has quickly learned to handle the hazards of Arctic diving—not only mechanical failures and physical privation, but the psychological stresses: the darkness that can swallow a diver in seconds, the fear of confinement under the icy expanse.

But we have no occasion to preen ourselves. Compared with the Eskimos, we are softies. Simon Idlout, who lives in a nearby Eskimo village, relates a tale that humbles us all. One day in spring he was crossing the ice near shore when his brand-new rifle dropped off his sled into the water. He went home for some rope, and, with his brother, returned to the site and stripped to his shorts. While the brother held one end of the rope, Simon wrapped the other around his waist, grabbed a large stone, and down he went some twenty feet. He stayed 50 seconds and did it again, looking around without benefit of goggles. He found the gun and trudged home to warm up. Obviously our friends in the North have much to teach us.

Doug and I now slip back into the water and make the long glide to *Sub-Igloo*. We tighten the last bolts and then begin to fill the eight-foot sphere with air. A huge bubble boils into *Sub-Igloo* and drives the displaced water down and through the open bottom hatch. As the air fills past the equator, we are relieved that there are no leaks. But occasionally I feel the structure tremble with tension; the potential lifting energy grows with each cubic foot of air entering the sphere.

The air is now down to the level of the circular plastic bench, well below the equator. It is time to go inside. I slip up through the hatch. My head breaks the surface. I remove my mask and hear the echo of my sigh. Cautiously I look around.

The water level is just below my chin, and I see small cakes of ice floating away on the wind of my steamy breath. The walls of *Sub-Igloo* seem not to exist.

I climb up on the bench and sit quietly. Three divers outside wave. One points beneath the bench. I lean over to look and see two small fish swimming there. I am in the Arctic's first undersea fishbowl. But the implications are larger than that.

The cold, clear water around me, so long the hidden home of arctic mammals and fish, is allowing us to probe its mysteries. This new tool, though, must be used with reverence. Yesterday an old Eskimo came into our tent, looked down the dive hole and said, "You will not scare our seals away, will you?"

Sub-Igloo is also, of course, a tool that has yet to prove itself fully. It is a prototype, and we must test it over long periods and in varied locations. But we now know that a device like this, which can be assembled so easily underwater, does work and will be useful.

In a few moments other divers will enter *Sub-Igloo*, and we will celebrate its successful Arctic christening with a champagne toast. Four of us will raise our glasses—after a battle with the stubborn cork. Awkward in our bulky gloves, we finally use a knife to pry the cork loose from its bottleneck.

In a few days I will experience another thrill—a 50,000-mile phone call from *Sub-Igloo* to Ottawa, to report our success to Prime Minister Pierre Elliott Trudeau, also a diver. Our words, carried by cable to Resolute, will be beamed from there to the new Anik I satellite—in orbit 23,000 miles above the earth's equator—and then bounced back to the Toronto area, to be relayed to Ottawa.

But for now I am alone, breathing resonantly inside the sounding sphere. And breathing, too, a silent thank-you. Our expedition has been a tremendous chain of planning and logistics. It began more than a year ago, 2,000 miles away in southern Canada, and it has involved aircraft, tracked vehicles, and men. Above all, men. Without my teammates' thousands of hours of effort, I would not be here, looking through this amazing window at the wonders of inner space.